Klaus H. Sames

Cryopreservation
Future Perspectives from Organ Transplantation to Cryonics

APPLIED HUMAN CRYOBIOLOGY

Edited by Klaus H. Sames
ISSN 2195-5700

1 *Klaus H. Sames (ed.)*
 Applied Cryobiology—Human Biostasis
 ISBN 978-3-8382-0458-1

2 *Klaus H. Sames (ed.)*
 Cryopreservation and Lifespan Extension
 Human and Animal Projects and Results
 ISBN 978-3-8382-0721-6

3 *Klaus H. Sames*
 Cryopreservation
 Future Perspectives from Organ Transplantation to Cryonics
 ISBN 978-3-8382-2058-1

Klaus H. Sames

CRYOPRESERVATION
Future Perspectives from Organ Transplantation to Cryonics

Bibliografische Information der Deutschen Nationalbibliothek
Die Deutsche Nationalbibliothek verzeichnet diese Publikation in der Deutschen Nationalbibliografie; detaillierte bibliografische Daten sind im Internet über http://dnb.d-nb.de abrufbar.

Bibliographic information published by the Deutsche Nationalbibliothek
The Deutsche Nationalbibliothek lists this publication in the Deutsche Nationalbibliografie; detailed bibliographic data are available on the Internet at http://dnb.d-nb.de.

The Cover illustration shows containers used in cryonics for the cryopreservation and storage of biological samples.
The term cryopreservation is used in this book for the preservation of biological entities by cooling them to temperatures below 0°C, preferably to cryogenic temperatures, and storing them at such temperatures.
Translated from "Kryokonservierung – Zukünftige Perspektiven von Organtransplantation bis Kryonik, Springer Spektrum 2022

ISBN (Print): 978-3-8382-2058-1
ISBN (E-Book [PDF]): 978-3-8382-8058-5
© *ibidem*-Verlag, Hannover • Stuttgart 2025

Leuschnerstraße 40
30457 Hannover
Germany / Deutschland
info@ibidem.eu

Alle Rechte vorbehalten

Das Werk einschließlich aller seiner Teile ist urheberrechtlich geschützt. Jede Verwertung außerhalb der engen Grenzen des Urheberrechtsgesetzes ist ohne Zustimmung des Verlages unzulässig und strafbar. Dies gilt insbesondere für Vervielfältigungen, Übersetzungen, Mikroverfilmungen und elektronische Speicherformen sowie die Einspeicherung und Verarbeitung in elektronischen Systemen.

All rights reserved. No part of this publication may be reproduced, stored in or introduced into a retrieval system, or transmitted, in any form, or by any means (electronic, mechanical, photocopying, recording or otherwise) without the prior written permission of the publisher. Any person who commits any unauthorized act in relation to this publication may be liable to criminal prosecution and civil claims for damages.

Printed in the EU

Table of Contents

Foreword 13
Preface 15

1 Introduction 17
2 How does cryonics work and what abilities are hidden in it? 21
3 Bridge to the future: distant goals of cryobiology and cryonics, unlimited times — unlimited possibilities 25
    3.1 A cryonics scenario 25
    3.2 Cryonics perspectives in medicine 26
        3.2.1. Deep cooling of the whole body 29
    3.3 Cryonics perspectives in spaceflight 29
    3.4 Cryonics perspectives in agriculture 31

4 The Physical Basis of Cryonics and the success of Related Methods 35
    4.1 Amazing benefits of the cold 40
    4.2. Thousands of years storage in liquid nitrogen 41
    4.3 Glass hard, pointed, physically and chemically dangerous: ice crystals 42
    4.4 Cryonics as a project design of modern biomedical research 43

5 Cryonics under the microscope — processes during cooling 49
    5.1 Deep-cooling of biological tissues 50
        5.1.1 Crystallization nuclei 50
    5.2 Diffusion, the migration of dissolved substances 53
        5.2.1 What happens during slow cooling of biological tissues below freezing point? 54
        5.2.2 What happens during rapid cooling? 55

5.2.3 Somewhere is the end, cooling makes the cell membrane tight 56
5.3 Survival of the cell on the narrow ridge of the cooling rate 57
5.4. Bone-hard liquid — vitrification 58

6 Tools for cryonics: Cryoprotectants — A big step forward but not yet the complete solution to our problems 67
6.1 Small but effective 69
6.2 What makes a cryoprotectant? 70
6.3 Large cryoprotectant molecules also have advantages .. 72
6.4 Molecule gathering up: individual cryoprotective agents and their effect 73
6.5 Side effects of cryoprotectants 79
6.6 Critical concentration changes of cryoprotectants 81
    6.6.1 What is actually damaged and how? 84
6.7 Increasing of the cryoprotectant concentration protection or death? Reduction of damaging effects Reduction of harmful effects 85

7 Other methods for protecting the cells and preventing ice crystals 95
7.1 Protection of vital membranes 95
7.2 Chemical fixation? 96
7.3 Methods to cope with huge amounts of data 97
7.4 How cryoprotectants help with glass formation (vitrification) 99
    7.4.1 Bound water 103
    7.4.2 Ice crystals are not tasty 104
7.5 Another advance: No more ice crystals? The ice blockers 105
    7.5.1 Too good to be true. 108
7.6 Clathrates: Water locks guest molecules in cages, ice must stay out — another chance? 109

- **8 Remaining hurdles and surprising solutions 117**
  - 8.1 How cells are fed, mass transfer between blood and cells 117
    - 8.1.1 Role of the cell membrane as the "mouth" of the cell 118
  - 8.2 Obstacles to the storage of cells and tissues 121
    - 8.2.1 Feasibility problems with vitrification and rewarming 121
  - 8.3. Problems with object size, chilling injury and cold shock 123
    - 8.3.1 Long distances, uneven structures 126
  - 8.4 Chilling injury and cold shock 128
    - 8.4.1 Living beings protect themselves 130
  - 8.5 Daring step to success: warming up 131
    - 8.5.1 Recrystallization, the comeback of ice crystals 131
    - 8.5.2 How warming leads to ice 132
    - 8.5.3 Does warming challenge the total advantages of deep cooling? 133
    - 8.5.4 Cryonics strategies to mitigate "devitrification" .. 134
  - 8.6 Premature use or experiment? Cryonics for prolonging the lives of people living today 136

- **9 A widely unknown success story: cryopreservation of cells embryos, tissues and small organs — many people have been "frozen" before 143**
  - 9.1 Cells 143
  - 9.2 Embryos 146
    - 9.2.1 Breakthrough in human reproduction 147
  - 9.3 Organs or parts of organs 148
    - 9.3.1 Tissue fractions from ovaries and testes 148
    - 9.3.2 Whole small organs 149
    - 9.3.3 Different sized organs, different temperatures from frosty to cryogenic 149

10 Important and promising starting points in the laboratory as well as in nature 169
10.1 Our irreplaceable biocomputer, the brain and nervous system 169
10.1.1 Nervous tissue 169
10.1.2 The Brain 170
10.2 A market for cryonics: artificially produced tissue and organs 172
10.3 An early success: cooling to freezing temperatures and reanimation 174
10.4 Nature a teacher for cryopreservation – is it better than our laboratories? 176
10.4.1 Hibernation 181

11 The body during oxygen deficiency 193
11.1 Ions and the electropotential of the normal, living cell 193
11.2 Without oxygen, cells work against themselves: Failure of the ion pumps 196
11.3 Exhausted batteries 198
11.4 Why is human resuscitation no longer possible after 9 minutes of cardiac arrest at the latest? 199
11.4.1 Fatal medical resuscitation: the reperfusion syndrome 199
11.4.2 Luck in misfortune for the cerebral cortex 205

12 Saving the human body: how is cryonics currently used to cryopreserve the human body? 211
12.1 The second worst thing that can happen to us: practice of cryopreservation a saving emergency procedure 211
12.2 Prerequisites for cryoprotectant perfusion 218
12.2.1 Influence of the carrier solution during perfusion 218
12.2.2 Effects of substances in a cell protection solution such as Viaspan 220

  12.2.3 Uptake of cryoprotectants from the blood into the cell 222

13 When taken over by cryonics: condition of a medically abandoned body 227
 13.1 When is a person dead? 227
  13.1.1 Life is still in it 227
  13.1.2 Biological definition of death and dying 228
  13.1.3 Evidence of death 229
 13.2 More than dead? How dead is finally dead? 230
  13.2.1 A small difference 230
  13.2.2 How do cells die? 232
  13.2.3 The sensational survival of nerve cells 234
  13.2.4 Resurrection of the pigs – survival of animal nerve cells 234
  13.2.5 Human nerve cells 239
  13.2.6 A suicide that lasts millions of years 241
  13.2.7 Not only nerve cells are masters of survival ... 242
 13.3 The new death 244

14 Cryonics can start to play along – intervention options after total organ failure 253
 14.1 Cooling still the best conservation method even after "death" 253
 14.2 Prevention of vascular occlusion 255
 14.3 Agents against cell death 258

15 Restoration 267
 15.1 Rewarming, recovery and reanimation 267
 15.2 Thawing and reanimation 268
 15.3 The finishing touch: Nano repair 270

16 Outlook: encouraging progress in cryonics research 277

Index 287

Dedicated to Ben Best, a renowned promoter of cryonics who provided me with inspiration and knowledge.

In Memoriam the "father of retina transplantation" and practicing cryonics proponent Peter Gouras and an unforgotten analytical mind and gifted gerontologist Bernhard L. Strehler.

Foreword

Cryopreservation of living objects is an evolving field of medicine and biology that offers decisive steps towards a revolutionary extension of medical methods.

This textbook is easy to read and describes the many different aspects of cryopreservation, the methods of cooling and rewarming, cryoprotection, and the study of freeze-resistant organisms in nature. The astonishingly high number of such organisms listed here is widely unknown.

The basic knowledge of cryopreservation is poor in the general public although there are an astonishing number of successful scientific experiments with cooling and rewarming of cells, tissues, organs and small organisms in different temperature ranges, facts that will visibly amaze the reader.

The book begins with an exciting description of the possibilities that would open up once cryopreservation and resuscitation of organisms were fully and easily developed and could be compared, for example, to anesthesia and reawakening, so that life could be stopped and restarted at will. Along those lines, one chapter deals with the state of preservation of the human body for minutes or a few hours after oxygen deprivation. It shows a sensational survival even of nerve cells of the brain during the observation period.

Importantly, the book provides a detailed observation of the behavior of biological tissues during slow or rapid cooling as well as vitrification at the cellular and molecular level. The role of a large number of cryoprotectants through their different properties and chemistry during such processes is presented.

It is also worth reading the book with regard to the preservation of the whole human organism (cryonics), describing in serious medical terms the opportunities and difficulties of cooling and rewarming patients who wish to be preserved after death to stop decay due to lack of oxygen as well as the cooling procedures themselves preserving the state after organ failure, where in many cases promising structural viability can be observed. The possibilities of future restoration of human bodies preserved and reanimated with new methods are assessed. Moreover, a professional assessment of today's cryonics methods is given. Their methods are presented realistically, their limitations, but also their opportunities are discussed.

The last chapter provides an outlook on the upcoming development of new methods. A fascinating extension of our ability to avoid damage during cooling and reheating, as well as to repair and initiate revitalization, is revealed. Medicine cannot afford to miss out on such progress. Furthermore, progress is also possible in space travel as well as, for example, in agriculture and the preservation of biodiversity.

Roman Bauer
Guildford (UK), November 2024

Preface

To stop aging, to rejuvenate their body and to prolong life is a dream goal of many people. However, through research and literary work, it became clear that the renewal of organs cannot be done by today's means, neither by drugs nor by gene therapy or stem cells. In principle, rejuvenation of our body is not impossible, but not in a short time frame. Aging and dying are still inescapable.

The only way available today to preserve the body of a deceased person is to cool it to -140°C or below, after introducing cryoprotectants into the blood circuit. We call this procedure cryonics. It seems a somewhat desperate measure, however, since no larger animal, and no warm-blooded animal at all, has yet been brought back to life from such temperature.

On the other hand, this is a thoroughly appealing field of research, not only because the seemingly impossible presents an intellectual challenge.

However, the preservation of biological entities by deep cooling (cryopreservation) opens up completely new possibilities for medicine and biology and associated technologies. The related field is cryobiology and where this targets the whole human being, we speak of cryonics.

Today we can already cool a little more gently than 10 years ago and that's what we do first and we do it with people in practice after death, because we have no other choice.

But perhaps we are laying the foundation for achieving an incredibly big goal.

The book intends these topics to become common knowledge excluding prepossession as well as blue-eyed or even utopic belief.

I want to thank my colleagues and friends for encouragement, support and stimulation: Peter Bezler, Paolo Brenner, Michael Dettmann, Matthias Erber, Benjamin Hampel, Wolfgang Krause †, Karlheinz Poch, Ramon Risco, Ludger Schmidt-Riese, Sebastian Sehte, Ralf Spindler, Alexandra Stolzing, Daniel Streidt and Robert Vöht.

My thanks for comfortable cooperation go in particular to Renate Scheddin and Dr. Sarah Koch from Springer Verlag, who advised me competently and purposefully, and to Mr. Padmanaban, who edited the text and Mr. Devarajan for production with technical expertise.

For production/publication of the English version I am indebted to Christian Schön publishing Director and Christoph Ohlwärther of *ibidem* publishers for advice and assistance.

Klaus Hermann Sames
Hersbruck, November 2024

1 Introduction

Summary

Today, when organs fail, medicine often has to give up.

However, there is an emerging possibility of preserving the body in an unaltered state rather than leaving it to decay. The limitation of medical options need not be final. Medicine continues to evolve. What is incurable today may be repairable tomorrow.

For long-term preservation of the body, there is only one currently applicable method: deep-cooling. We name the procedures concerned "cryopreservation" they will be described in chapters of this book. Biological units (for example transplant organs) could be preserved by this method, because cooling stops chemical processes. When applied to humans, we speak of cryonics. The methods are not yet fully developed, but cryonics is without alternative. Scientists are working intensively on methods to cure diseases, stop aging processes and rejuvenate. Those alive today will only be able to participate in this if their bodies are preserved in as unaltered a state as possible when their organs fail. This book is intended to show the current effectiveness of procedures concerned.

The preservation of life is the dominant task of medicine, but aging and diseases cannot be stopped definitively today.

This could change with a method that stops the decay of the body even after the circulation stops, such as cryonics.

Today, cryonics can only be carried out using methods that are not yet perfect. First of all, flow of cryoprotectants through the blood circuit and other complicated measures are necessary. We will deal with the circulation of organs in various ways and discuss the methods of cryonics in detail below.

Cryonics can at least be seen as an opportunity to maintain the body in a state of potential resuscitation through deep freezing.

It is interesting to ask whether cryonics will not be superfluous one day due to gerontological methods such as stopping the aging processes, regenerative medicine and rejuvenation — or will become pure emergency medicine.

Can we now stop time our own time (measured by changes of our body) by deep-freezing? Or how could we personally enjoy the possible success of gerontology in 100 years or even later?

Deep freezing is actually a method that stops the changes in the body as far as that is possible. However, the body is then also not alive and that cannot be what we want in the long run. In each case, we want to keep the body in a functional state.

Of interest to us is only a technology that can be used immediately, i.e. also for people living today, and does not rule out resuscitation, even if it is perhaps not yet fully developed and thus contains a major risk.

A significant impact on aging or even rejuvenation for the generations living now is not yet possible, even if the media and scientific portals are buzzing with promises and premature announcements. The problems can only be solved over a long period of time. A renewal of all components of the body in their special structure and function is not yet conceivable.

As long as rejuvenation methods do not work on the living, they can only be postponed to the future. All people must still die for the foreseeable future.

However, to the best of our knowledge, there is no fundamental obstacle to life extension or rejuvenation. Repair of the body could be effective mainly with the help of gene therapy or stem cell manipulation, but with certainty only by nanotechnology.

In order to preserve a body even after the organs have failed until then, it must be saved from final decay, dying must be interrupted death must be postponed, biological time must be stopped.

Cryonics is not a particularly elegant option, but it is the only one that is already within reach.

We hope that a restoration of the structures and sufficient energy will also make the functions of the organs possible again, just as in surgery the heart recovers after bloodlessness and a standstill of function when it is warmed up and receives blood again. Like a car that runs again after repair and refueling. Whether the car has a soul can be debated if one is so inclined. It does not (yet) possess a self-repair like a living organ.

It must be emphasized: Cryonics today represents the only possible way to significantly prolong our lives and reach an age of more than 123 years. For this purpose, the body must be preserved after organ failure and pronouncement of death, whether it is old or young, sick or healthy at the time of failure of life processes.

Other methods are conceivable, but not yet ready for application (see also Fischer R. 2019).

Cryonics is a young method that still has a long development ahead of it. However, this development could be exciting.

Cryonics advocates hope that methods can be used to restore a body that has been cryonically stored even if it has various damages.

The failure of all our organs ("death"), lack of oxygen, aging and diseases cause more or less severe damage, which today is only partially analyzed and cannot be repaired.

So, resuscitation and healing are still left to the future.

Cryonics is often mistakenly referred to as the freezing of bodies after death. Through cryonics there would then be a resurrection.

Of course, human beings do not have the resurrection from the dead in their hands. However, we can partially preserve what is still viable and interrupt the dying process.

Cryonics is not about corpses, because the purpose of freezing is to preserve living cells and structures of the body as well as possible — no more, no less.

Freezing is not the right term either, rather we intend crystal-free deep-cooling that allows cells and tissues to survive.

Literature

Fischer R (2019) Sterben war gestern – ein praktischer Ratgeber ISBN 978-3-9820734-0-8 (www.sterben-war-gestern.de)

2 How does cryonics work and what abilities are hidden in it?

Summary

The future-oriented cryonics project is defined. Its goal is described. Realistic steps towards its realization are envisaged and the present situation is outlined.

Cryonics is the name given to the preservation of living beings at very low temperatures and their intended rewarming and resuscitation when these include humans as the center. It must from the beginning also include all living beings and biological materials existing to develop its methodology.

It also involves an application of methods that are still immature, but which are currently the best according to the state of the art (reviews on cryonics: Best 2008; -2013; Bojic et al. 2021; Fahy 2021; Fahy, Wowk 2021; Mathwig, Sames 2013; Reinhard 1987; Sames 2011a; Sames 2013a; -b; -2018; Sames et al. 2022).

Cryonics can only prolong human life by interrupting the dying process. It does not have a direct effect on stopping aging or on rejuvenation. Today, among all life-sustaining methods, the most perfect deep-cooling is undoubtedly a priority, while other problems are still reserved for research.

Cryonics aims to store the human body unchanged until medicine is able to heal the damage caused by aging, disease, death and by the cooling processes themselves. The success of cryonics cannot (yet) be guaranteed.

In contracts, it is even legally essential that the patient takes note: neither future resuscitation nor funding into a more distant future can be guaranteed.

The principle of cryonics is biostasis. This means a reduction or, strictly speaking, a standstill of life processes, for example, by desiccation (aestivation) or by lowering the temperature deep below zero (cryostasis) or near zero but still above (hibernation). All states of biostasis should allow a revival of the living being.

Cryobiology and cryonics do not use biostasis by desiccation. They rely on cryostasis (deep temperature biostasis), which is the only method that already works in the laboratory today.

Since the resuscitation of mammals after deep cooling is not yet possible, cryonics on living humans cannot be allowed today. Death must be awaited and with it usually the consequences of aging, causes of death and dying, the most important obstacles to prolonging life. The inspection of the corpse must also be completed, so that in Germany the start of cooling is often only allowed after two hours, which runs counter to the concept of preservation. However, the damage caused by long periods without cooling storage could be significantly reduced if the inspection for cryonics candidates would be performed in cold rooms. After early inspection of the head, it should be cooled with ice if there are no findings on the head. At most, virtopsy (virtual biopsy) a form of dissection with only the smallest of interventions should be used before cryonics, if already possible and urgently needed. The rapid development of these methods is to be demanded on the part of cryonics.

Aside from the personal desire of people alive today for self-preservation, cryonics is an experimental scientific project with a very broad goal.

But what about the possibilities that would be offered by functioning cryonics with people and their organs being deep-cooled, stored in the cold, and resuscitated?

Literature

Best BP (2008) Scientific justification of cryonics practice. Rejuvenation Res 11:493-503

Best B (2013) Cryonics: Introduction and technical challenges. In Sames KH (ed) Applied Human Cryobiology, vol. 1., ibidem, Stuttgart, pp 61-77

Bojic S et al (2021) Winter is coming: the future of cryopreservation. BMC Biol 19, 56

Fahy GM (2021) Principles of vitrification as a method of cryopreservation in reproductive biology and medicine. In Donnez J, Kim SS (eds) Fertility preservation principles and practice, section 2-Reproductive biology and cryobiology. Cambridge University Press, pp 35-66

Fahy GM, Wowk B (2021) Principles of Ice-Free Cryopreservation by Vitrification. Methods Mol Biol 2180:27-97, doi: 10.1007/978-1-0716-0783-1_2

Mathwig K, Sames K (2013) Kryonik. In Sun MJ, Kabus A (eds) Reader zum Transhumanismus. Books on Demand, Norderstedt, Berlin, pp 113-129

Reinhard K (1987) Wie der Mensch den Tod besiegt. Technische Verfahren zur Unsterblichkeit. Orac, Wien

Sames KH (2011) "wollt ihr ewig leben?" durch die Kryonik zum ewigen Leben? Innsbrucker Forum für Intensivmedizin und Pflege (IFIMP) Univ. Klinik für Allgemeine und Chirurgische Intensivmedizin. Medizinische Universität Innsbruck 09-10 Juni http://www.intensiv-innsbruck.at/meetings/IFIMP2011_bilder.htm

Sames KH (2013a) General mechanisms of mortality and aging and their relation to cryonics. In Sames KH (ed) Applied Human Cryobiology, vol. 1, ibidem, Stuttgart, pp 145-169

Sames KH (ed) (2013b) Applied Human Cryobiology, vol. 1, ibidem, Stuttgart

Sames KH (ed) (2018) Applied Human Cryobiology, vol. 2, ibidem, Stuttgart

Sames KH et al (2022) Safe preservation of organs by cryogenic cooling, a chance of the future not only of medicine. In Willmann TA, El Maleq A (eds) Sterben 2.0 (Trans-)Humanistische Perspektiven zwischen Cyberspace, Mind Uploading und Kryonik. De Gruyter, Basel, pp 221-238

3 Bridge to the future: distant goals of cryobiology and cryonics, unlimited times — unlimited possibilities

Summary

One may let the imagination wander to come across remote possibilities. What opportunities would be functioning cryonics open up for us? How could space travel, medicine or agriculture benefit and what would be possible after a human's organs have failed? In almost naively simplistic optimistic scenarios, an attempt is made to explore what possibilities lie in the cryonics project and where it already works today.

3.1 A cryonics scenario

A short skit is designed to show, in 8 short scenes, the excitement of saving a patient through cryonics.

1. The heart stops, the blood is no longer supplied with oxygen by the lungs.
2. the cells get no more oxygen and run out of energy.
3. But they are still alive, while the human being is already lying motionless and unconscious with missing brain waves and usually does not come back to life. The cells try to exploit reserves like energy-rich phosphates — or creatinine compounds and sugars still trying to gain energy to run their molecular pumps.
4. Finally, the chemical energy reserves are also completely depleted. The cells resemble empty batteries.
5. the molecules now change, damaging substances are formed and those that are chemically reduced. Fortunately, the blood no longer flows, because it increasingly contains harmful waste products and it is over acidified.
6. that's when cooling sets in!
   The body has fallen into cold water or a medical or cryonics representative applies artificial cooling.

The cells then go to sleep. Their metabolism becomes so slow that they hardly need any more energy. They also produce hardly any waste products. Cooling has occurred in time to stop the greatest damage. Many cells are still in a state in which they could be revived.
7. antifreeze agents ensure that ice crystals do not form as far as possible. Cells are additionally protected by drugs.
8. if the body is now cooled further to -196°C, the temperature of liquid nitrogen, today's cooling methods will cause additional damage, but the components of the body will last for an unimaginably long time afterwards, as long as the temperature remains at -196°C.

One thing is certain: many cells were not yet dead. According to all we know, their form and former function will still be recognizable in a few thousand years.

That's a lot of time for medicine to develop methods for the patient's repair and recovery.

We want to wish him luck.

3.2 Cryonics perspectives in medicine

But cryonics is much more than that. The aim is no more and no less than to switch life on and off at will. It could be used to suspend and restart life at any time and for any duration (if fully functional).

Thus, a revolutionary protection and rescue method is also aimed at. In deep cooling damage can be stopped and "frozen". A protection against destructive effects can be built up, since a frozen body has no needs, no food, no excretion, no air, it can be put into resistant protective covers, maybe even cast in concrete. The only thing it really needs is protection against heating. If cryonics were to work fully, it would be a rescue method for medicine as well, far superior to those used today. It could even protect humans against extreme environmental conditions. A person could become almost invulnerable through it.

An important step in development of cryonics is represented by preservation of tissues and whole organs in transplantation medicine to create organ banks of cryopreserved organs for transplantation to humans.

Recently, artificial tissues have also been cryopreserved with initial success. Cryonics here should meet the needs of an emerging large market. Since cryonics candidates cannot donate their organs but must keep them for themselves, they are very interested in the development of artificial organs.

Cryopreservation of transplant organs would represent an enormous medical advance, apart from the fact that it would come close to cryopreservation of the human body.

There are far too few donor organs available for transplantation medicine today, due to difficulties in preserving and transporting organs. According to the WHO, in 2010 only ten percent of the donor organs needed worldwide were available. In 2017 (all organs combined) 139,024 organs were transplanted (again, only 10% of the world's need) and, more importantly, appropriate storage methods are lacking. Despite this, an estimated 20% of donor kidneys, up to 50% of donor pancreas glands, and about two-thirds of donor hearts are discarded in the United States primarily because of the time required for donor search and transport. Many are transferred to unsuitable recipients in the rush that is required (Ardehali 2015; Global Observatory on Donation and Transplantation GODT; Guibert et al. 2011; Ibrahim et al. 2020; Parsons et al. 2014; Reese et al. 2016; Taking Organ Utilization to 2020: WHO 2012).

Usually, organ preservation is performed in immobile cooled liquid of 4-10°C (hypothermic preservation) or the organ is perfused with various solutions. Known solutions are e.g.: University of Wisconsin, Euro Collins or Celsior solution (Bessems et al. 2005; Wolkers, Oldenhof 2021). Cold storage quickly leads to cold damage and blood deficiency damage or damage upon reperfusion. This also causes a high proportion of donor organs to be discarded. This high rate of loss could be avoided if organs could be preserved for a longer period of time, and more transplantations would be possible. Thousands of organs are discarded due to the short organ survival time alone (Israni et al. 2017).

Millions of lives worldwide would also benefit from being able to replace organs and tissues on demand rather than under time constraints (Giwa et al. 2017). If cryopreservation of large organs, including their resuscitation, were to be realized, one could stock an assortment of donor organs in organ banks to reduce waiting times for patients. One could also transport donor organs without harm or time pressure (and without helicopters). No more organs would become unusable purely because of time constraints. It is claimed that this would totally solve the shortage of e.g. transplant hearts.

Even hours of additional preservation would increase the number of possible transplantations. Some extension of organ survival is allowed by continuous perfusion that occurs at body temperature (35-37°C) during survival or a combination of perfusion with lowered temperature of 4-10°C, as routinely used in kidney transplantation. It can also extend lung survival from 8 to 21 hours (Yeung et al. 2017). The addition of cryoprotectants and suppression of ice crystal formation allowed blood vessels to perfuse at even lower temperatures (-6°C, hypothermia) in rat livers. This extended storage to 4 days (Berendsen et al. 2014). Thus, methods like those intended by cryonics are already in development in other medical areas (e.g. heart surgery).

However, today there are still few effective methods for reliable preservation of human transplant organs for longer than 3-12 hours (Bruinsma et al. 2014; Simpkins et al. 2007; Totsuka et al. 2002).

In turn, many medical interventions would benefit from cryopreservation (Israni et al. 2017; Lewis et al. 2016; Taylor et al. 2019).

To study the conditions for preservation, one does not necessarily need whole human organs. Tissue sections and artificial tissues closely resemble the organ from which they were taken and contain all cell types found in the living organ (Chen et al. 2018; de Graf et al. 2007; Li et al. 2016; Pichugin et al. 2006).

In this way, it would also be possible to use human tissue samples in research, which would always be available in consistent quality thanks to deep-cooling. Furthermore, testing drugs and toxins animal testing could be more easily avoided (de Graaf et al. 2003; Kramer, Greek 1990; Li et al. 2016; Sandow et al. 2015; Truskey 2018; Zaman et al. 2007). Costs could be saved in expensive drug development (as costs to bring a drug from the laboratory to the market exceed EUR 2 billion (DiMasi et al. 2016; review: Bojic et al. 2021).

This would also bring us a little closer to the final step of cooling a person deeply.

3.2.1. Deep cooling of the whole body

Preserving the entire human body by cooling it below 0°C would logically give medical professionals much more time to intervene.

Overall, fully developed cryonics would then also allow doctors to keep a patient, they can no longer keep alive, in his current state in order to search for new remedies. For example, it could be used to save a patient who is bleeding from numerous vessels so that doctors cannot keep up with stopping the bleeding. It would be similar with life-threatening organ damage. Children who die of cancer could be frozen if there is a prospect that a cure for the cancer will be available in a few years. Patients could survive cryopreserved until a new drug or cure is developed. Transporting an otherwise untransportable patient to an out-of-town specialty hospital, for example, would be possible. Perhaps a doctor could simply cryonize his patient when he is at a loss. The thorny issue of fatal malpractice should also be mentioned here, because doctors are not infallible. A doctor would not have to give up on a patient he loses, no matter what the cause of organ failure. He could keep him in the state of organ failure and thus buy time and look for new ways.

As I said, all this applies in the event that cryonics works and we overcome all the hurdles that still stand in its way. The first thing to note here is that it is worth it.

3.3 Cryonics perspectives in spaceflight

Sophisticated cryopreservation would be a big win for space travel (Nordeen, Martin 2019). The idea is so common that it already seems banal.

How broad the conceivable possibilities of cryonics are becoming clear when one asks how a human being on Venus could survive in a heat hell if he crashes there, for example, or urgently needs to get to a certain point in person. A super-insulating mechanically super-strong capsule—preferably with its own cooling unit—would perhaps be one possibility, resting life processes another. We leave the further fate of the deep-frozen space traveler to the imagination.

You could transport by transport bring large groups of people to other planets, for example, and store them until they are all there to establish a colony.

Or let's ask how to make spaceships smaller and easier to supply. Space journeys to other solar systems, which take centuries, could be managed as generation flights or more simply with the help of cryonics. Only with the help of cryonics would a space traveler taking off from Earth be able to personally reach such a far destination. The space organizations are definitely engaged in such planning strategies (see below).

Currently, some of the biggest obstacles to long-duration travel (to Mars, for example) have to do with the needs for sustaining the lives of space travelers. One needs resources to maintain metabolism and bodily functions. There are significant health hazards from prolonged exposure to interplanetary radiation and weightlessness, as well as psychostress from spatial confinement, leading to sociological problems. Cryopreservation could make these problems insignificant, in part because it allows total physical and psychological shielding of a person. The European Space Agency (ESA) and the U.S. National Aeronautics and Space Administration (NASA) have therefore been investigating the possibility of suspended animation for deep space flights. (Ayre et al. 2004; Choukèr et al. 2019; Petit et al. 2018; "Torpor Inducing Transfer Habitat for Human Stasis to Mars" https://ntrs.nasa.gov/citations/20180008683; "Advancing Torpor Inducing Transfer Habitats for Human Stasis to Mars" https://ntrs.nasa.gov/citations/20180007195; Szondy D: ESA studies impact of hibernating astronauts on space missions: https://www.esa.int/Science_Exploration/Human_and_Robotic_Exploration/Exploration/European_vision_of_exploration).

3.4 Cryonics perspectives in agriculture

For agriculture, it is already possible to store genetic samples or cell cultures of many plants and animals frozen today (Barboni et al. 2018; Sponenberg 2020)

But what if whole animals could be frozen and brought back to life, whether to take them out of the market for a time or to transport and stockpile them safely and gently?

Animals could be kept viable frozen and stacked in animal banks. Here, too, assortments of large quantities of different animal and plant species could be kept in stock, if one does not prefer to transport only the DNA and clone the living beings, which, however, is also costly and much more time-consuming, or to acquire frozen embryos, which one then still has to bring to development. Imagine acquiring a frozen hamster at the pet store, then thawing it according to instructions once the husbandry conditions have been established. Human cryonics could also benefit from this through price reduction and new research opportunities.

Agriculture, like medicine, may one day wonder why it did not promote cryonics from the beginning.

The whole thing could have the side effect of easily moving from large animal cryonics to human cryonics. If the ongoing experiments were also successful on transplant organs and eventually on large animals and became applicable to humans, cryonics could also be used to treat healthy individuals who are less affected by aging and disease. Death would no longer have to be waited for. One could intervene before we suffer irretrievable damage due to aging, disease, or accidents, e.g., by "freezing" young space travelers. We hope that one day the stress of cryonization will not be greater than that of anesthesia. This would make the rescue method fully applicable.

All in all, many different ethical aspects speak for the fact that it would be an unforgivable omission not to try the development of cryonics. The responsibility for this must already be assumed today by those who promote this omission out of prejudice or insufficient information. Above all, one can expect so much insight from medicine that it actively promotes the development of such a promising method.

One could live a very long life by cryonics, an unlimited life, but not—as far as at all foreseeable—an immortal one.

How close are we actually to the revolutionary goals of cryonics? We will go into detail below. Here it may be stated that their impossibility cannot be asserted so far.

Literature

Ardehali A (2015) While millions and millions of lives have been saved, organ transplantation still faces massive problems after 50years; organ preservation is a big part of the solution. Cryobiology 71:164-165

Ayre M et al (2004) Morpheus—Hypometabolic Stasis in Humans for Long term Space Flight. J Br Interplanet Soc 57:325-339.

Barboni B et al (2018) Placental stem cells from domestic animals: translational potential and clinical relevance. Cell Transplant 27:93-116

Berendsen TA et al (2014) Supercooling enables long-term transplant survival following 4 days of liver preservation. Nat Med 20:790-793

Bessems M et al (2005) Improved machine perfusion preservation of the non-heart-beating donor rat liver using polysol: A new machine perfusion preservation solution. Liver Transplant 11:1379-1388

Bruinsma BG et al (2014) Subnormothermic machine perfusion for ex vivo preservation and recovery of the human liver for transplantation. Am J Transplant 14:1400-1409

Chen R et al (2018) A study of cryogenic tissue-engineered liver slices in calcium alginate gel for drug testing. Cryobiology 82:1-7

Choukèr A (2019) Hibernating astronauts-science or fiction? Ploughman's Arch 471:819-828

De Graaf IA, Koster HJ (2003) Cryopreservation of precision-cut tissue slices for application in drug metabolism research. Toxicol In Vitro 17:1-17

De Graaf IA et al (2007) Cryopreservation of rat precision-cut liver and kidney slices by rapid freezing and vitrification. Cryobiology 54:1-12

DiMasi JA et al (2016) Innovation in the pharmaceutical industry: new estimates of R&D costs. J Health Econ 47:20-33

Giwa S et al (2017) The promise of organ and tissue preservation to transform medicine. Nat Biotechnol 35:530-542

Global Observatory on Donation and Transplantation GODT www.transplant-observatory.org

Guibert EE et al (2011) Organ preservation: current concepts and new strategies for the next decade. Transfus Med Hemother 38:125-142

Ibrahim M et al (2020) An international comparison of deceased donor kidney utilization: what can the United States and the United Kingdom learn from each other? Am J Transplant 20:1309-1322

Israni AK et al (2017) OPTN/SRTR 2015 Annual Data Report: deceased organ donation. Am J Transplant 17 Suppl 1:503-542.

Kramer LA, Greek R (1990) Human stakeholders and the use of animals in drug endothelium-dependent vasodilatory responses in canine coronary arteries following cryopreservation. Cryobiology 22:511-520

Lewis JK et al (2016) The grand challenges of organ banking: proceedings from the first global summit on complex tissue cryopreservation. Cryobiology 72:169-182

Li M et al (2016) Precision-cut intestinal slices: alternative model for drug transport, metabolism, and toxicology research. Expert Opin Drug Metab Toxicol 12:175-190

Nordeen CA, Martin SL (2019) Engineering human stasis for long-duration spaceflight. Physiology (Bethesda) 34:101-111

Parsons RF, Guarrera JV (2014) Preservation solutions for static cold storage of abdominal allografts: which is best? Curr Opin Organ Transplant 19:100-107

Petit G et al (2018) Hibernation and Torpor: Prospects for Human Spaceflight. In Seedhouse E, Shayler D (eds) Handbook of Life Support Systems for Spacecraft and Extraterrestrial Habitats. Springer, Cham, p 1-15

Pichugin Y (2006) Cryopreservation of rat hippocampal slices by vitrification. Cryobiology 52:228-240

Reese PP et al (2016) new solutions to reduce discard of kidneys donated for transplantation. J Am Soc Nephrol (JASN) 27:973-980

Sandow N et al (2015) Drug resistance in cortical and hippocampal slices from resected tissue of epilepsy patients: no significant impact of P-glycoprotein and multidrug resistance-associated proteins. Front Neurol 18 February 2015, https://doi.org/10.3389/fneur.2015.00030

Simpkins CE et al (2007) Cold ischemia time and allograft outcomes in live donor renal transplantation: is live donor organ transport feasible? Am J Transplant 7:99-107

Sponenberg DP (2020) Conserving the Genetic Diversity of Domesticated Livestock (editorial). Diversity 12:282

Taking Organ Utilisation to 2020, https://www.odt.nhs.uk/odt-structures-and-standards/key-strategies/taking-organ-utilisation-to-2020/

Taylor MJ et al (2019) New approaches to cryopreservation of cells, tissues, and organs. Transfus Med Hemother 46:197-215

Totsuka E et al (2002) Influence of cold ischemia time and graft transport distance on postoperative outcome in human liver transplantation. Surg Today 32:792-799

Truskey GA (2018) Development and application of human skeletal muscle microphysiological systems. Lab Chip 18:3061-3073

WHO (2012) Keeping kidneys. Bull World Health Organ 290:718-719

Wolkers WF, Oldenhof H (2021) Principles underlying cryopreservation and freeze-drying of cells and tissues. Methods Mol Biol 2180:3-25

Yeung JC et al (2017) Outcomes after transplantation of lungs preserved for more than 12 h: a retrospective study. Lancet Respir Med 5:119-124

Zaman GJ et al (2007) Cryopreserved cells facilitate cell-based drug discovery. Drug Discov Today 12:521-526

4 The Physical Basis of Cryonics and the success of Related Methods

Summary

Cold is not always hostile to life. It can preserve biological objects and thus human bodies unchanged for thousands of years, but it also poses dangers. One must know the processes involved in cooling in order to control cooling and also to use it medically. How to deal with ice crystals that damage living cells? Does a plan exist?

Cryobiology is much older than it may seem to us today. As early as the 18th century, Reaumur conducted scientific observations and found that some insects survive the winter in a frozen state (see in Block et al. 1990). We will return to such insects below.

It has also been known for a long time that people who drown in cold water are occasionally still revived after an hour or more. Medicine is therefore already making use of cooling in the form of hypothermia.

In hypothermia, the body temperature can be lowered by 10-20°C with the administration of cell protection drugs. This allows the heartbeat to be stopped for a longer period of time without harm to the patient.

In fact, cold has been used in medicine for centuries to treat fever, reduce blood loss, for anesthesia, and (unconsciously) to delay cell death.

Hypothermia was already known to the ancient Egyptians, Chinese, Greeks and Romans. Hippocrates advised wrapping severely injured people in ice and snow to reduce blood loss (Zorn 2013).

Among the early experimentalists of the 17th century, Boyle is particularly noteworthy. He noted the ability of ice to preserve human bodies and made various experiments to freeze and revive animals. In the process, he discovered species of frogs and fish that could survive confinement in ice (Boyle 1665).

Larrey, Napoleon's personal physician noted, similar to ancient observers, that wounded survived better and were in less pain if they lay in the snow away from the fire.

In 1952, Lewis and Taufic operated on the heart under refrigeration. A heart-lung machine was developed and first used on the open heart of the gibbon in 1953.

The cooling form of deep hypothermia enabled procedures on the heart under circulatory arrest. Artificial circulation outside the body under moderate hypothermia is standard in cardiac surgery. Excess potassium can be used to stop the heart. Survival rates are highest with rapid cooling (Alam et al. 2002; -2004). A modern cooler in the oxygenator cools 5-6L of blood from 36 to 18°C in 20 min. Modern surgical methods allow operating in mild hypothermia (32-34°C). For example, a venous cooling catheter was inserted via the femoral vein with advancement into the inferior vena cava, close to the heart, cooled with an external cooler to 33°C for 24 hours, and then passively rewarmed over 4-5 hours.

Patients with cardiac arrest have been found to have better brain scores with the use of mild hypothermia (Hypothermia after Cardiac Arrest Study Group 2002; Nozari et al. 2004) and the number of survivors is found to be greater after hypothermia than without hypothermia (Holzer et al. 2002; Nolan et al. 2003; - 2010; Von Lewinski, Pieske 2011).

In emergency medicine, mild hypothermia has been used in cardiac arrest, i.e., when blood flow has stopped, but usually after successful resuscitation before transport to a hospital. However, immediate cooling is probably better.

Reducing temperature from 37°C to 32-34°C for 24 hours after cardiac arrest in humans reduced brain damage and increased the number of survivors. After 6 months, 55% more survivors were noted (Cheung 2006; Couzin 2007; Holzer 2002).

The so-called deep hypothermia allows cooling to 18°C and a blood flow stop of about one hour in human patients. Hypothermia thus led to good successes (Lyon et al. 2010; Zorn 2013).

If the brain is perfused separately, cooling no longer plays such a decisive role today. Brain damage, however, still represents the most serious side effects of surgery.

Experiments on gerbils have shown that when the temperature is lowered from 37 to 31°C, nerve cells survive a lack of fresh blood better. The time for which nerve cells endure blood stasis is almost tripled (Takeda 2003).

Temperature reduction also inhibited the effects of harmful oxidation products — oxidative stress — and the release of transmitter substances caused by potassium in brain cells in mice and rats (Khandoga et al. 2003; Kimelberg et al. 1995; Kollmar et al. 2002).

Cooling above 0°C (hypothermia) also reduces damage to the blood-brain barrier (see below) by inhibiting enzymes that break down protein (Hamann et al. 2004; Raison 1973). The first few degrees of temperature reduction reduce the formation of inflammatory substances (cytokines) in the brain to one third compared to that at normal body temperature (Meybohm et al. 2010).

Cooling could also slow the formation of blood clots (thrombi) from plasma, fibrin fibers, and blood cells after cardiac arrest.

In gerbils, a temperature decreases from 37- to 31°C nearly triples the time after which neurons can still recover from a blood flow arrest (Takeda et al. 2003).

Dogs endure 60 min of circulatory arrest at 20°C body temperature, compared with 120 min at 10°C (Behringer et al. 2003).

In mice, temperatures below 15°C significantly reduce oxidative stress (Kandola et al. 2003).

Cooling stops changes and can thus be used in the worst case even after the organs have failed. Nothing else happens in cryonics as a first measure.

Today, the vast majority of corpses in Germany are also refrigerated, but bacterial growth is inevitable.

Although cooling is the best means against cell decay, it does not protect absolutely (Vollbracht et al. 2001). Unfortunately, despite cooling in the range above 0°C, tissue damage still occurs during cold blood flow arrest (cold ischemia). In the case of renewed perfusion or perfusion with solutions after a cold blood flow stop, different changes are found compared to the stop of blood flow in warmth (warm ischemia).

> Background information
> The enzyme for the formation of nitric oxide is inhibited and eicosanoid is formed, which constricts the blood vessels. Chelatable iron appears and this is able to open certain pores (MPTP) on the respiratory organelles at such low temperatures, which can lead to cell suicide or rapid cell death (Hansen et al. 2000; Rauen et al. 2004).

Other damage can also occur at low temperatures above freezing. For example, chilling injury and cold shock must also be considered above 0°C (Al-Fageeh, Smales 2006; Hays et al. 2001).

A surprising discovery led one step further. If the blood is replaced by a cold isotonic solution — i.e. total oxygen deprivation with simultaneous cooling — cardiac arrest can be survived at even significantly lower temperatures. Such a state could last up to 3 hours at about 10°C (i.e. far below the temperature limit survivable normally by mammals).

These facts show that more highly evolved mammals, probably including humans, can be cooled below the critical limit of 15°C.

The method has now reached clinical applicability and is expected to be used soon (Alam et al. 2002; Behringer et al. 2003; Bellamy et al. 1996; Bulger et al. 2010; Drabek et al. 2007; Kheirbek et al. 2009; Kutcher et al. 2016; Liu et al. 2017; Rhee et al. 2000; Safar et al. 2000; Schreiber et al. 2015; Tisherman SA 2017; Wu et al. 2006).

> Background information
> In one experiment, the initial washout of the blood was done with a solution corresponding to the tissue fluid outside the cells. It was later replaced by a cell-type solution with low levels of sodium ions and high levels of potassium ions. After heating, the first solution was used again. Then the blood was returned (Taylor 1995; -2019).

S. Tisherman told New Scientist (Health 20 November 2019) that this method called: "emergency preservation and resuscitation" (EPR, formerly "suspended animation") has been used in at least one human case and that he would report scientifically on all cases of his project in 2020.

A similar condition has been induced in mice and pigs by hydrogen sulfide (H2S) in the blood (Blackstone et. al 2005; Li et al 2008; Morrison et al 2008; Roth, Nystul 2005; Simon et al 2008), but not in large animals such as sheep (Haouzi et al 2008). Maassen and coworkers (2019) reduced the metabolism of pig kidneys with H2S. As a result, oxygen consumption decreased by 61%. Remarkably, chemical energy stores (esp. ATP) and kidney structure and function were preserved.

Liu et al (2017) did not completely remove the blood from minipigs. The heart was stopped and perfused with cold organ protection solution and blood via an artificial circuit. It was cooled to 15°C. All animals survived 90 minutes of cardiac arrest. At 120 minutes of cardiac arrest, 2 animals survived and 5 died after 2 days. Thus, they did not survive significantly longer than patients undergoing cardiac surgery in deep hypothermia.

Roth and Nystul 2005 reported on interesting experiments on oxygen deprivation. Larvae of the nematode C. elegans also fall into a dormant state during radical oxygen deprivation, which is actively induced by enzymes. However, if some residual oxygen remains, then the animals die. It could be that the residual oxygen allows the formation of harmful products, especially free radicals. Whether this is also the case in suspended animation of mammals remains open. This could mean that a washout of blood by an oxygen-free solution in cryonics has a similar beneficial effect on cells, but is compromised by residual oxygen. The washout would have to be performed very early.

A step toward lower temperatures could be supercooled without crystal formation (for more details see Taylor et al. 2019). A refrigerator with an electrostatic field even allows small organs to be stored in this state (Monzen et al. 2005).

These are sensational advances in the cooling of warm-blooded organisms, but we are still far from zero with the temperature and thus even further from the realm of cryonics namely cryogenic temperatures. In order to preserve cells and biological tissues for a long time, temperatures must be lowered far below freezing.

4.1 Amazing benefits of the cold

Cryonics is a field of thermobiology, i.e. the heat balance of a body, but at temperatures deep below 0°C, which do not occur in the nature of the earth. However, a general principle of each temperature reduction is the throttling of chemical processes, thus also of the metabolism. Lowering metabolism by cooling naturally decreases the cell's demand, making it less dependent on blood flow (Hayashida et al. 2007). During cooling, the cells consume fewer and fewer nutrients and oxygen for metabolic processes down to temperatures at which simple chemical metabolic processes require millennia.

Incidentally, it is not surprising that temperature also has an influence on quite normal life processes. Even without cryonics, it plays a decisive role in metabolism and aging processes.

In cold-blooded animals, low ambient temperature may prolong life by reducing metabolism (see Lamb 1977).

The body temperature of mice can be lowered by a specific diet, with a concomitant increase in lifespan (Weindruch, Walford 1988).

Hibernators have a longer lifespan than comparable nonhibernators.

Whether these findings can be used for cryonics remains to be investigated.

For living objects, a rule of thumb (at least above 0°C) is that a 10°C decrease in temperature decreases metabolism by 50% (Belzer 1988).

Medicine is moving towards cryonics, so to speak: from higher to lower temperature levels of hypothermia and currently towards deep freezing.

Chilled perfluorocarbon has been attempted to be used for accelerated cooling either delivered through the lungs or through the blood vessels. This substance can also serve to thin the blood at normal body temperature without damaging the small blood vessels — such as capillaries. The special feature is that it supplies oxygen. Therefore, it can be delivered to the lungs (even during life) (Cabrales 2004, Dinkelmann, Northoff 2000).

4.2. Thousands of years storage in liquid nitrogen

It can be theoretically calculated that at very low temperatures, normal metabolic processes largely come to a standstill.

At the temperature of liquid nitrogen (boiling point -196°C), the heat-related energy for chemical reactions is absent and only radiation (e.g., cosmic background radiation) causes damage. Hits are rare, but accumulate in inactive tissue over tens of thousands of years. Simple metabolic processes theoretically last up to 10 million years at nitrogen temperature (Karlsson, Toner 1996).

In addition, the water reaches a solid state, through which chemical processes are hindered. This implies the possibility of a very long practically unchanged storage of the remains of a man. Arrhenius' equation $k = A \exp(-Ea/RT)$ shows a relationship between the speed of chemical processes and temperature (here Ea is the activation energy, R the gas constant and A a frequency factor for molecular collisions (see Best 2008; Karlsson, Toner 1996; Mazur 1984).

It is fascinating to imagine that one's own body can be preserved for such a long time, and perhaps possibilities for the future are still open to a person.

Even just below -130 °C, the viscosity of solutions is so high (greater than 1013 poise) that molecular motion is practically insignificant. In practice, no cell death was found in liquid nitrogen for 2-15 years, and even when irradiated with several times 100 the background radiation for 5 years, the same is true. Even more interesting: Mouse embryos at the 8 cell stage were exposed to radiation equivalent to 2000 years of background radiation in liquid nitrogen for 5-8 months. The animals developed from this exposure later showed no discernible consequences of deleterious effects either in their survival or in their development (Glenister et al. 1984; Mazur 1963; -1984).

The enzyme acetylcholinesterase shows shape changes upon X-ray irradiation at -118°C but no longer at -173°C (Weik et al. 2001).

On enzymes, it has also been shown that temperatures below that of liquid nitrogen reduce even X-ray damage (Chinte et al. 2007; Meents et al. 2010).

However, there is also reason for concern. Lettuce seeds showed some damage after 10-20 years of storage at nitrogen temperature. A graph was made showing that if you measure the time at low temperatures after which 50% of the seeds no longer germinate, this period becomes as longer as the temperature drops. At -135°C this was 500 years, and at -196°C it was 3400 (Walters et al. 2004). Molecular movements in the reading of DNA in cells have been suggested as an explanation for the decrease in seed quality at the low temperatures (Wowk 2010).

> Background information
> Unfortunately, the so-called quadratic vibration of water molecules is not completely inhibited even at -273°C (with 0.0171 square angstrom) (Leadbetter 1965)

4.3 Glass hard, pointed, physically and chemically dangerous: ice crystals

The possibility of preserving living organisms, including humans, at extremely low temperatures is tempting. However, getting to low temperatures is not easy for living entities and we need to study the processes that occur during cooling in detail if we want to preserve tissues and organs through cooling. Cryonics cannot be done on the fly, but requires in-depth knowledge.

At the freezing point already lurks the greatest danger with which cryonics is confronted. The crystallization of water is capable of destroying cells mechanically, chemically and osmotically (Adam et al. 1990; Mazur 1963; Pegg 2010). We will discuss this in more detail.

4.4 Cryonics as a project design of modern biomedical research

The application of cryobiology and cryonics is gradually moving from simple, very small creatures to more complicated and larger creatures or parts of creatures.

Cooling and resuscitation of cells and tiny tissue samples is now routine.

In the last decades the same succeeded with small organs, but only in the case of the quite small kidneys of the rabbit or rat as well as the ovary of the mouse so perfectly that animals could live normally with the thawed kidney and the ovaries were fertile. We will come back to this.

The next step, which is already underway, is to find a method that all organs can tolerate to cryopreserve an entire small animal — such as a mouse — and bring it back to life.

The next problem would then be the deep-cooling of larger organs, for example for banks of human transplant organs.

Eventually, larger animals could be envisioned.

The critical leap then is to the living human. For deep cooling and preservation at very low temperatures (cryopreservation) and resuscitation, the experiences in deep cooling of deceased and in the clinical application of hypothermia, as well as those in infarcts and experiments on animals and biological tissues, can provide valuable guidance. The first and most important disciplines using cooling when blood flow stopped were cardiac surgery and neurosurgery, which learned to work on the arrested heart or brain. From them, the methods for cryonics have been largely adopted and there is close collaboration.

In any case, we should already develop concrete ideas about how to proceed with large living objects.

Literature

Adam M et al (1990) The effect of liquid nitrogen submersion on cryopreserved human heart valves. Cryobiology 27:605-614

Alam HB et al (2002) Learning and memory is preserved after induced asanguineous hyperkalemic hypothermic arrest in a swine model of traumatic exsanguination. Surgery 132:278-288

Alam HB et al (2004) The rate of induction of hypothermic arrest determines the outcome in a Swine model of lethal hemorrhage. J Trauma 57:961-969

Al-Fageeh MB, Smales CM (2006) Control and regulation of the cellular responses to cold shock: the responses in yeast and mammalian systems. Biochem J 397:247-259

Behringer W et al (2003) Survival without brain damage after clinical death of 60-120 minutes in dogs using suspended animation by profound hypothermia. Crit Care Med 31:1523-1531

Bellamy R et al (1996) Suspended animation for delayed resuscitation. Crit Care Med. 24 (2 Suppl), p 24-47

Belzer FO, Southard JH (1988) Principles of solid organ preservation by cold storage. Transplantation 45:673-676

Best BP (2008) Scientific justification of cryonics practice. Rejuvenation Res 11:493-503

Blackstone EA et al (2005) H2S induces a suspended animation-like state in mice. Science 308 (5721):518

Block W, Baust JG et al (1990) Cold tolerance of insects and other arthropods [and Discussion]. Phil Trans Roy Soc Series B, Biological Sciences 326:613-633

Boyle R (1665) New experiments and observations touching cold. J. Crooke, London

Bulger EM et al (2010) ROC Investigators: out-of-hospital hypertonic resuscitation following severe traumatic brain injury: a randomized controlled trial. JAMA 304:1455-1464

Cabrales P et al (2004) Oxygen delivery and consumption in the microcirculation after extreme hemodilution with perfluorocarbons. Am J Physiol Heart Circ Physiol 287:H320-H330

Cheung KW et al (2006) Systematic review of randomized controlled trials of therapeutic hypothermia as a neuroprotectant in post cardiac arrest patients. CJEM 8:329-337

Chinte U et al (2007) Cryogenic (<20 K) helium cooling mitigates radiation damage to protein crystals. Acta Crystallogr D Biol Crystallogr 63:486-492

Couzin J (2007) The Big Chill. Science 317:743-745

Dinkelmann S, Northoff H (2000) Lebenserhaltung durch künstliche Sauerstoffträger. In Sames K (ed) Medizinische Regeneration und Tissue Engineering, neue Techniken der Erhaltung und Erneuerung von Gewebefunktionen. Ecomed, Landsberg, VII-6, S 1-3

Drabek T et al (2007) Emergency preservation and delayed resuscitation allows normal recovery after exsanguination cardiac arrest in rats: a feasibility trial. Crit Care Med 35:532-537

Glenister PH et al (1984) Further studies on the effect of radiation during the storage of frozen 8-cell mouse embryos at -196 degrees C. J Reprod Fertil 70:229-234

Hamann GF et al (2004) Mild to moderate hypothermia prevents microvascular basal lamina antigen loss in experimental focal cerebral ischemia. Stroke 35:764-769

Hansen TN et al (2000) Warm and cold ischemia result in different mechanisms of injury to the coronary vasculature during reperfusion of rat hearts. Transplant Proc 32:151-158

Hayashida M et al (2007) Effects of deep hypothermic circulatory arrest with retrograde cerebral perfusion on electroencephalographic bispectral index and suppression ratio. J Cardiothorac Vasc Anesth 21:61-67

Hays LM et al (2001) Factors affecting leakage of trapped solutes from phospholipid vesicles during thermotropic phase transitions. Cryobiology 42:88-102

Haouzi P et al (2008) H2S induced hypometabolism in mice is absent in sedated sheep. Respir Physiol Neurobiol 160:109-15

Holzer M et al (2002) Mild therapeutic hypothermia to improve the neurologic outcome after cardiac arrest. N Eng J Med 346:549-556

Karlsson JEM, Toner M (1996) Long-term storage of tissues by cryopreservation: critical issues. Biomaterials 17:243-256

Khandoga A et al (2003) Impact of intraischemic temperature on oxidative stress during hepatic reperfusion. Free Radic Biol Med 35:901-909

Kheirbek T et al (2009) Hypothermia in bleeding trauma: a friend or a foe? Scand J Trauma Resusc Emerg Med 17:Article 65

Kimelberg HK et al (1995) Astrocytic swelling due to hypotonic or high K+ medium causes inhibition of glutamate and aspartate uptake and increases their release. J Cereb Blood Flow Metab 15:409-416

Kollmar R et al (2002) Neuroprotective effect of delayed moderate hypothermia after focal cerebral ischemia: an MRI study. Stroke 233:1899-1904

Kutcher ME et al (2016) Emergency preservation and resuscitation for cardiac arrest from trauma. Int J Surg 33:209-212

Lamb MJ (1977) Biology of aging. Blackie, Glasgow

Leadbetter AJ (1965) The thermodynamic and vibrational properties of H2O ice and D2O ice. Proc Royal Soc A 287:403-425, https://doi.org/10.1098/rspa.1965.0187

Lewis FJ, Taufic M (1953) Closure of atrial septal defects with the aid of hypothermia; experimental accomplishments and the report of one successful case. Surgery 33:52-59

Li J et al (2008) Effect of inhaled hydrogen sulfide on metabolic responses in anesthetized, paralyzed, and mechanically ventilated piglets. Pediatr Crit Care Med 9:110-112

Liu Y et al (2017) A safety evaluation of profound hypothermia-induced suspended animation for delayed resuscitation at 90 or 120 min. Mil Med Res 4:16

Lyon RM et al (2010) Therapeutic hypothermia in the emergency department following out-of-hospital cardiac arrest. Emerg Med J 27:418-423

Maassen H et al (2019) Hydrogen sulphide-induced hypometabolism in human-sized porcine kidneys PLoS One 14:e0225152

Mazur P (1963) Kinetic of water loss from cells at subzero temperatures and the likelihood of intracellular freezing. J Gen Physiol 47:347-369

Mazur P (1984) Freezing of living cells: mechanisms and implications. Amer J Physiol 247 (3 Pt 1) C 125-142

Meents A et al (2010) Origin and temperature dependence of radiation damage in biological samples at cryogenic temperatures. Proc Natl Acad Sci U S A 107:1094-1099

Meybohm P et al (2010) Mild hypothermia alone or in combination with anesthetic post-conditioning reduces expression of inflammatory cytokines in the cerebral cortex of pigs after cardiopulmonary resuscitation. Crit Care 14:R21

Monzen K et al (2005) The use of a supercooling refrigerator improves the preservation of organ grafts. Biochem Biophys Res Commun 337:534-539

Morrison ML et al (2008) Surviving blood loss using hydrogen sulfide. J Trauma 65:183-188

Nolan JP et al (2003) Therapeutic hypothermia after cardiac arrest. An advisory statement by the Advancement Life Support Task Force of the International Liaison Committee on Resuscitation. Circulation 108:118-121

Nolan JP et al (2010) European Resuscitation Council Guidelines for Resuscitation 2010. Section 1: Executive Summary. Resuscitation 81:1219-1276

Nozari A et al (2004) Mild hypothermia during prolonged cardiopulmonary cerebral resuscitation increases conscious survival in dogs. Crit Care Med 32:2110-2116

Pegg DE (2010) The relevance of ice crystal formation for the cryopreservation of tissues and organs. Cryobiology 60 (3 Suppl):36-44

Raison JK (1973) The influence of temperature-induced phase changes on the kinetics of respiratory and other membrane-associated enzyme systems. J Bioenerg 4:285-309

Rauen U et al (2004) Iron-induced mitochondrial permeability transition in cultured hepatocytes. J Hepatol 40:607-615

Rhee P et al (2000) Induced hypothermia during emergency department thoracotomy: an animal model. J Trauma 48:439-447; discussion 447-50

Roth MB, Nystul TG (2005) Survival in cryogenic sleep. Spectrum Wiss, Sept:42-48

Safar P et al (2000) Suspended animation for delayed resuscitation from prolonged cardiac arrest that is unresuscitable by standard cardiopulmonary-cerebral resuscitation. Crit Care Med 28 (11 Suppl): N214-N218

Schreiber MA et al (2015) A controlled resuscitation strategy is feasible and safe in hypotensive trauma patients: results of a prospective randomized pilot trial. J Trauma Acute Care Surg 78:687-697

Simon F et al (2008) Hemodynamic and metabolic effects of hydrogen sulfide during porcine ischemia/reperfusion injury. Shock 30:359-364

Takeda Y et al (2003) Quantitative evaluation of the neuroprotective effects of hypothermia ranging from 34°C to 31°C on brain ischemia in gerbils and determination of the mechanism of neuroprotection. Crit Care Med 31:255-260

Taylor MJ et al (1995) A new solution for life without blood. Asanguineous low-flow perfusion of a whole-body perfusate during 3 hours of cardiac arrest and profound hypothermia. Circulation 91:431-444

Taylor MJ et al (2019) New approaches to cryopreservation of cells, tissues, and organs. Transfus Med Hemother 46:197-215

Tisherman SA et al (2017) Development of the emergency preservation and resuscitation for cardiac arrest from trauma clinical trial. J Trauma Acute Care Surg. 83:803-809

Volbracht C et al (2001) Apoptosis in caspase-inhibited neurons. Mol Med 7:36-48

VonLewinski D, Pieske B (2011) "Cooling" after cardiopulmonary resuscitation in 2011. Intensivmed 48:185-189

Walters C et al (2004) Longevity of cryogenically stored seeds. Cryobiology 48:229-244

Weik M et al (2001) Specific protein dynamics near the solvent glass transition assayed by radiation-induced structural changes. Protein Sci 10:1953-1961

Weindruch RL, Walford RL (1988) The retardation of aging and disease by dietary restriction. Charles C Thomas, Springfield Illinois

Wowk B (2010) Thermodynamic aspects of vitrification. Cryobiology 60:11-22

Wu X et al (2006) Induction of profound hypothermia for emergency preservation and resuscitation allows intact survival after cardiac arrest resulting from prolonged lethal hemorrhage and trauma in dogs. Circulation 113:1974-1982

Zorn H (2013) The ice age in cardiac surgery and rescue medicine. In Sames KH (ed) Applied Human Cryobiology, vol. 1, ibidem, Stuttgart, pp 171-179

5 Cryonics under the microscope — processes during cooling

Summary

The processes involved in cooling are observed in this chapter. The formation of ice crystals is temperature-dependent and requires nuclei. The question of how to prevent them has long occupied the scientific community. What happens in biological tissues during cooling and ice crystal formation is the key to avoiding hazards and maintaining the viability of frozen tissues.

When reading biomedical research results, one must always be aware of one difficulty of research. It is difficult to study cells within living tissues without harming them. It is also difficult to study a tissue made up of many cells, but which is only a tiny part of an organ, and it is again difficult to study an organ as a part of a large body. On the other hand, it is also difficult to infer from a person's behavior how their liver is doing. So you can't study a body at all levels of cells, tissues and organs at the same time. For the most part, studies are carried out at intervals from one another and on units artificially separated from the body, such as individual cells, pieces of tissues or total organs. Conversely, they are performed on the whole body without looking into every cell.

Many reactions and effects are tested on cells, for example. However, one cannot simply transfer such results to organs or the whole human being. Thus, from the reaction of a culture of isolated liver cells, one cannot necessarily infer the reaction of the whole liver, and from the reaction of a liver, one cannot always infer the overall condition of a human being. Rather, one must test an intact liver under the same conditions as the liver cells and, finally, the values in a whole living body that indicate the work of the liver. Eventually, medicine is based on uniting the processes at all levels into an overall picture.

The cooling and reheating of cells, tissues, organs and whole living or failing bodies must practically try anew to form such a holistic picture, which exists only partially in this field.

For example, studies on the behavior of liver cells during freezing cannot simply be equated with the behavior of the whole liver, or substances that affect the metabolism of the liver can simply be tested during the cooling of cell cultures and not also on the whole liver. Unfortunately, reports do not always mention exactly how the results were obtained. Often, of course, the results on cells are confirmed by the results on their organs. but there may be also serious differences due to the structure of the organs comparing biological values determined in cells with those determined in organs. However, the effects of substances and cooling methods in particular are often described only on the basis of studies on cell cultures. As far as possible, we mention in the following for which examination material and which procedure the results apply, but this cannot always be done here either. It would become a boring enumeration of methods. Therefore, when reading, one must always critically question how far the validity of an investigation goes and what was actually investigated. In case of doubt, one must consult the sources.

Now, why is it not yet possible to freeze and thaw people today in order to bring them back to life?

5.1 Deep-cooling of biological tissues

We are not powerless against the devastating formation of ice crystals in human tissues. We can analyze the complicated processes involved in crystallization and not only that, we can intervene and even control the processes.

5.1.1 Crystallization nuclei

The initial formation of crystals, during freezing in an aqueous sample is only possible with the help of crystallization nuclei. The growth of crystals follows afterwards (secondary nucleation). Various theories have been formed about these complicated processes (Jones 2002; Karthika et al. 2016; Kashchiev 2000; Vali 1995).

Although the melting temperature of water is 0°C, totally pure water will not freeze at temperatures above -40°C. At such temperatures, water requires crystallization nuclei to trigger crystal growth (Lundheim 2002).

Crystallization nuclei are also represented by the ice crystals themselves—so to speak own nuclei of the water (homonuclei) in contrast to also effective external particles (heteronuclei). Tap water contains enough external crystallization nuclei to crystallize at 0°C (e.g. in freezers). If there are few crystallization nuclei, however, water can be supercooled. For example, it remains liquid down to -40°C. However, if the first ice crystals form in the supercooled water, they spread explosively, since ice crystals themselves serve as crystallization nuclei (Karlsson et al. 1993). The temperature of ice nucleation (nucleation temperature) depends on chance (it occurs stochastically).

The formation of ice crystals endangers living cells. It is therefore useful to control nucleation (Petersen et al. 2006; Vali 1996). The temperature at which even pure water freezes (-40°C) is called the homogeneous (uniform) nucleation temperature, because the nuclei themselves are of the same substance as ice crystals.

For the growth of crystals in pure water to start, however, a certain amount of water must first be present, the critical amount of water. At -5°C, this critical amount of water is 45,000 molecules of water. At -20°C only 650 molecules are needed and at -40°C only 70, a tiny amount (Vali 1995). With the method of vibrational spectroscopy, crystallization can even be made visible (Pradzynski et al. 2012).

The higher the temperatures are above -40°C, the lower the tendency for ice to form.

Not surprisingly, in animals, the likelihood of ice formation increases with body size (more precisely, the volume or weight of the body).

Thus, some species of reptiles with a body mass of less than 20 grams can tolerate hypothermia below -5°C (Costanzo, Lee 1995; Costanzo et al. 1995).

In a supercooled biological sample, the tissue fluid can be induced to form ice crystals by touching it with a cooled needle (induction). Crystallization can be induced by this or by administering crystallization nuclei if one wishes to determine the time or temperature of crystal formation.

For large specimens (where a pinprick does too little), cooling is performed to the desired ice formation temperature, and then cooling is accelerated until ice is formed on the surface of the specimen. Crystallization then progresses from the surface to the depth. With rapid cooling of such larger tissue samples, the temperature inside remains at the ice formation temperature. Only the surface temperature initially decreases further (for methods see Karlsson, Toner 1996).

Several other methods exist for actively controlling the nucleation process during slow freezing (Dalvi-Isfahan et al. 2017; Morris, Acton 2013). The temperature at which ice nuclei form can be controlled by creating localized supercooled conditions. These are created by cooled metal rods, expanded gases, ultrasound or laser pulses, silver iodide, nanoparticles, or the like, among others. The probability of formation of crystallization nuclei starting can also be increased by applying an electric field. In addition, there are ice nucleating proteins that have been discovered in various frost-resistant organisms such as bacteria (Anastassopoulos 2006; Braslavsky, Lipson 1998; Chow et al 2003; Han et al 2007; König et al 1997; Lindinger et al 2007; Lundheim 2002; Margaritis, Bassi 1991; Morris et al 2003; Petersen et al 2003; Zacchariassen, Kristiansen 2000 cited following Bojic et al 2021).

In solutions such as those present in biological tissues, in addition to the presence of crystallization nuclei, all types of solutes play a role in the formation of ice crystals. These processes have been studied directly in lowering the temperature in tissues. Thereby the rate of cooling is important.

Deep cooling for live preservation (cryopreservation) may be considered routine today for many cell types in cell cultures. 90% of the cells or more survive such a procedure for many cell strains.

Also, many tissues surprisingly endure a transformation of more than 80% of their water into ice crystals. Even in most whole organs, no damage occurs when 40% of their water is present as ice (see also Best B Vitrification in cryonics. https://www.benbest.com/cryonics/vitrify.html).

5.2 Diffusion, the migration of dissolved substances

Aqueous solutions inside living cells are separated from solutions outside the cells by the cell membrane. The membrane has a limited permeability (it is semipermeable). The permeability is not the same for all substances. Some pass through the cell membrane almost unhindered, others only slowly, and still others not at all. If a substance is strongly enriched in the cell, it migrates (diffuses) – if possible – through the cell membrane into the tissue fluid. Conversely, if a substance is more accumulated in the tissue than in the cell, it may migrate through the cell membrane into the cell to equalize the concentration. The greater the difference in concentration, the faster the molecules migrate (Fick's law of diffusion). Small molecules generally migrate faster across the cell membrane. Increased temperature accelerates the passage (diffusion).

If the migration of the atoms or molecules is not possible – for example, because they do not pass through the cell membrane due to their size – or if a balance has not yet been achieved, a high concentration arises on one side of the membrane which builds up a pressure, the osmotic pressure against the other side where the concentration is lower.

A solution in the body can have the same osmotic pressure as the blood or cell fluid. The different solutions are then isotonic. In a solution that has a higher osmotic pressure than the solutions in the cell, the cell shrinks. A sudden change in osmotic pressure can therefore damage cells. Cells can withstand highly enriched solutions in which they shrink. If molecules enter the cell from the outside and draw water after them, the cell swells again. This is called a shrink/swell cycle. Cells are more sensitive to re-swelling than to shrinking (Paynter 1999 cited following Best B Perfusion & Diffusion in Cryonics Protocol. http://www.benbest.com/cryonics/protocol.html).

Bodies contain many solutions with the same osmotic pressure, so no problems arise.

5.2.1 What happens during slow cooling of biological tissues below freezing point?

Normally, a rate of about 1-50°C/min is used for slow cooling (Jang et al. 2017). However, different cells have different optimal cooling rates. Water transport across the cell membrane changes with temperature (Mazur 1963). The concentration of cryoprotectant used is sometimes 2.5-10% and a maximum of 1.5 M. Diffusion and osmotic pressure are critical for maintaining tissues during cooling. In general, the interior of the cell contains more solutes than the environment of the cell, but less external nuclei. Therefore, the formation of ice crystals occurs first outside the cells (see also Best B Physical parameters of cooling in cryonics, http://www.benbest.com/cryo nics/cooling.html).

Ice crystal formation inside the cell is more dangerous, because ice crystals have about 10% larger volume than liquid water and this expansion can damage the cell (Petrenko, Whitwort 1999). In the process, a cell can even burst.

Slow cooling of biological tissues now results in the binding of water in the form of ice outside the cells. The removal of liquid water leads to solution thickening (increase of concentration) in the remaining residual liquid. This then forces an outflow of water from the cells. The cells shrink until the concentrations equalize. Unfortunately, ice in the space adjacent to the cell can damage the cell (Wowk 2007).

Such processes therefore put a strain on the cell. However, there is also a thickening of the solutions in the cell when water flows out, which favorably reduces the formation of ice crystals inside the cells (Karlsson, Toner 1996). This is because over-concentrating of substances lowers the freezing point. These then act like cryoprotectants. In addition, this reduces the expansion of the water during ice formation.

However, the removal of water from the cell may eventually reach a critical point (critical minimum cell volume or -residual volume). If it falls below this, cell shrinkage can also be fatal for the cell (Meryman 1970). For animal cells, the critical limit is 30-50% water loss. It has been shown that too rapid dehydration damages cell membranes. This also alters proteins, nucleic acids, and fat bodies (liposomes) (Wolkers et al. 2007; Wolkers, Oldenhof 2021). An accurate time course of crystallization can prevent water loss from exceeding this limit (Muldrew et al. 2004). However, at the depth of large tissue districts, it is difficult to estimate the time required because cooling arrives there more or less delayed. Thus, one has to try to modulate shrinkage in a different way, e.g., by using substances that hinder ice formation outside the cells or substances that retain water inside the cells.

With slow cooling, the escape of water from the cell—which takes time—is high. In this case, the high concentration of intracellular solutions allows crystallization inside the cell at lower temperatures if at all at most. As supercooled water leaves the cell, a so-called thermodynamic equilibrium can form. In this case, the water is bound in the form of ice outside the cell. Ideally, the interior of the cell can undergo vitrification, i.e. solidification without crystallization (Wowk, Fahy 2007 and see below).

5.2.2 What happens during rapid cooling?

If cooling is too fast, ice can form inside the cell. Again, ice formation begins outside the cells and the concentration of the unfrozen liquid is increased and water is drawn out of the cells. However, because the water cannot leave the cell quickly enough, the fluid in the cells remains poorly concentrated and ice crystals can form with further cooling (Fuller et al. 2004; Wesley-Smith et al. 2015; Yu 2017). A lethal amount of ice in cells has been determined to be 3.7% in cultured liver cells (Karlsson et al. 1993).

The formation of ice crystals inside the cell during rapid cooling is more dangerous than ice formation outside the cell and, as mentioned, can be fatal to it (see in Karlsson, Toner 1996; Lynch, Diller 1981; Sputtek 1996).

> Background information
> To make it a bit more complicated, crystal formation within the cell itself also binds liquid water, so it also leads to increased concentration of the still-liquid solution and thus slows crystal formation

somewhat down. Additionally rapid cooling allows less time for crystallization to occur, and this can also hamper crystallization.

Thus, the speed of cooling or cooling rate alone already has a decisive influence on the condition that a cooled tissue reaches in the end. It even makes a difference whether one cools at a steady rate or the cooling rates change during cooling. This results in very elaborate experiments for different ways of proceeding in cryonics (see Karlsson, Toner 1996). One can try to determine the most favorable conditions in each case. Today, the use of computer programs is emerging for this purpose.

It is very important that large bodies can be cooled and heated much more slowly than small ones. Ultimately, the option of cooling and revival of whole people also depends on optimization of such methods.

5.2.3 Somewhere is the end, cooling makes the cell membrane tight

At a certain temperature around -40°C, the leakage of water from the cell ceases, no matter how fast you have cooled and what the concentration of substances is. This is due to the permeability of the various membranes that protect the cell and its organelles. It decreases as the temperature drops. Cell membranes undergo phase transitions during cryopreservation (Spindler et al. 2011). Membranes do not have a dimensionally stable structure. They are also not structureless like water, but their well-ordered molecules allow for viscous flow. For the cell, changes in this viscosity of membranes with temperature change are a problem. The behavior of cell membranes upon temperature change, their reaction with cryoprotective agents, and the manipulation of their permeability (e.g., to channel cryoprotectants into the cell), have been presented in more detail by Wolkers and Oldenhof (2021).

5.3 Survival of the cell on the narrow ridge of the cooling rate

There exists an optimal cooling rate at which both ice formation in the cell and water removal from the cell with cell shrinkage are low. It lies in the middle between very high cooling rate (retention of water and ice formation in the cell) and very low cooling rate (water loss and shrinkage). Their knowledge allows a more targeted approach (Karlsson, Toner 1996).

This favorable cooling rate can unfortunately be orders of magnitude different for different cell types. It depends on the properties of the cell membrane and the volume, and also on the type of cryoprotectant and its concentration.

For cryopreservation of tissues, organs or even whole living beings this is a serious problem, while isolated cells can be cooled by exact rates (in the right area).

A computer model was created to determine cell survival at a given cooling rate (Bauer 2021).

At the moment the only thing left to do is to analyze damage in the experiment and look for a method that causes the least damage for all cell types contained in a tissue.

To see how damaging ice crystals are, one needs to know what they do to the cell. Ice crystal formation can damage cells in ways other than shrinkage, e.g., mechanically (Adam et al. 1990). Ice crystals outside the cell can deform cells. Cells can also be spatially constricted by an increasing amount of solidifying fluid in their environment. However, crystals do not grow through the cell membrane (Acker et al. 2001).

Fortunately, ice formation within the cell is not harmful in itself if the extent remains small. However, it could set mechanical damage especially to membranes, because a direct adhesion between the cell membrane and ice is also supposed to cause damage. The formation of gas bubbles by ice inside the cell is another possibility of damage.

When the solutions thicken, when ice crystals bind the water to themselves, fat-protein compounds can change (literature in

Karlsson, Toner 1996; Steponkus et al. 1983). Unfortunately, the vital membranes are made of such compounds.

Another possibility of damage during rapid cooling is membrane rupture due to strong osmotic flow of substances and water through the membranes

The result of different types of cooling was studied by: 1. immersion in liquid nitrogen 2. a more gradual temperature reduction to -130°C. 3. cooling in a freezer (to -130°C), which also served for storage.

Ice crystal formation was the same for all three cooling modes.

It is very important that ice crystal formation in biological tissue prefers zones with reduced flow in blood vessels. This was shown using ink.

It means that you have to make sure that really all parts of an organ are flowed through during cooling (de Wolf, de Wolf 2013).

During melting, water is released again and the concentrations change. Such changes during melting of the ice are therefore also a possible cause of damage (Mazur et al. 1972; Muldrew, McGann 1994).

According to what has been said, it is clear that the formation of ice crystals and the osmotic load of cells are main obstacles for cryonics. However, they are not insurmountable. An improved way to prevent them is vitrification, which we will get acquainted with next.

5.4. Bone-hard liquid — vitrification

In 1937, Luyet vitrified biological material for the first time.

In a glass, the molecules are randomly oriented, as in a liquid. Glass can flow. This is why very old church windows are said to become thicker at the bottom, as you can read in guidebooks. In reality, however, it may just be a matter of known manufacturing defects. Window glass flows at room temperature over such long periods of time that we cannot yet observe it (Plumb 1989; Zanotto 1998). Whether the glass also flows in the organs of cooled, vitrified patients has not been definitively determined. In any case, it is not yet possible to vitrify entire human organs without any ice crystals.

In general, the demonstration of glass within cells and tissues is very complicated and requires at least a cryo-electron microscope (Hübinger 2013). Due to the disordered molecules, a glass has little structure and this has the advantage of being transparent. However, the advantage of vitrification for cryonics is that it avoids the damage caused by structure formation in the form of ice crystals (e.g., Armitage 2002; Fahy 2021).

With rapid cooling (e.g. 100°C per minute), ice formation can be avoided if glass formation occurs instead. At a certain temperature, the glass transition temperature, there is a rapid increase in viscosity with solidification of the water or solution with no increase in volume as would be observed with crystallization. Thus, another hazard disappears (McFarlane 1987).

Glass formation is mainly achieved by rapid cooling, which, so to speak, gives no chance to the formation of crystals. The freezing point is undercut. Undercooling of the liquid occurs (see Armitage 2002). With further cooling, the viscosity increases until solidification occurs, unless crystallization occurs first.

One advantage is that the cooling rate no longer has to be set specifically to the cell type. The disadvantages of fast cooling (see above) during vitrification are accepted.

The production of glass from pure water is complicated. Since water has a low viscosity and low thermal conductivity, the vitrification of pure water can only be achieved by a shock-like reduction in temperature below the freezing point. Pure water only vitrifies during cooling by millions of °C per second to -135°C. At such rapid cooling, even the water molecules do not have enough time to arrange themselves into crystals (see also Best B vitrification in cryonics. https://www.benbest.com/cryonics/vitrify.html).

During freezing, samples — e.g. solutions with cells — normally form ice first in the edge zone, where the cooling arrives first. This ice can itself stimulate further crystallization. The release of latent heat, which is always released during crystallization, increases the temperature lag between the edge temperature and the internal temperature. As a result, the cooling rate in the interior will be even higher than in the edge zone. For vitrification of high-volume specimens, the cooling must be slow enough to achieve a uniform distribution of temperature, especially in the glass transition temperature region. This can ideally avoid stresses and cracking, but unfortunately the speed cannot be changed arbitrarily (Steif et al. 2007).

For glass formation, the layer thickness of the biological tissue is thus important, because it largely determines how quickly or slowly the object as a whole cools or heats up (Hochi et al. 2001). The cooling rate and heating rate both play a major role. For example, Kuleshova and coworkers (1999) used a cooling rate of 100°C per minute and a heating rate of 10°C per minute to vitrify a 59 percent ethylene glycol salt solution.

By avoiding ice crystals, vitrification techniques made it possible for the first time to obtain images of biological tissues that showed no changes even at high magnification with an electron microscope.

The glass transition temperature can change. It depends on how fast cooling is used and which solution is used. With faster cooling, the glass formation is already reached at higher temperatures (increase of the glass transition temperature). This is one way to control vitrification.

Vitrification techniques subsequently made it possible to bring organs such as the small rabbit kidney and the tiny ovary (ovary) of mice back to life from cryogenic temperatures.

But even with the conventional method, a number of (even larger) organs survived sub-zero temperatures.

Further progress could be made with directional freezing.

In the usual freezing process, the heat generated during crystallization (heat of fusion) is conducted through the frozen portion of the sample. This can trigger melting processes resulting in cell damage. The microscopic changes and the rate of ice formation are uncontrolled and not uniform.

In directional freezing, heat is dissipated through the unfrozen portion of the sample. This is achieved, for example, by cooling from one side so that a heat gradient is created from the cooled to the uncooled side. The directional ice growth leads to the formation of lamellae that grow from the cooled side toward the uncooled side and between which the cells are trapped, reducing their mechanical damage (Arav 2022; Saragusty 2015).

Directional cooling also allows to avoid ice formation within cells during faster cooling (Bahari et al. 2018).

A related technique the directional vitrification using the same principle but with very high cooling rate is in development. This technique could achieve vitrification with concentrations of cryoprotectant as low as 17.5% (Arav, Natan 2012) providing a major step forward.

Both methods should be used to lower the temperature to -130°C (Bojic et al. 2021; Mazur 1984).

The control possibilities that the cooling rate offers us are amazing. However, they are not yet sufficient to reliably prevent crystal formation in a human body.

Literature

Acker JP et al (2001) Intracellular ice propagation. Experimental evidence for ice growth through membrane pores. Biophys J 81:1389-1397

Adam M et al (1990) The effect of liquid nitrogen submersion on cryopreserved human heart valves. Cryobiology 27:605-614

Anastassopoulos E (2006) Agar plate freezing assay for the in situ selection of transformed ice nucleating bacteria. Cryobiology 53:276-278

Arav A, Natan D (2012) Directional freezing of reproductive cells and organs. Reprod Domest Anim (47 Suppl) 4:193-196

Arav A (2022) Cryopreservation by Directional Freezing and Vitrification Focusing on Large Tissues and Organs. Cells 11:1072

Armitage WJ (2002) Recovery of endothelial function after vitrification of cornea at -110 degrees C. Invest Ophthalmol Vis Sci 43:2160-2164

Bahari L et al (2018) Directional freezing for the cryopreservation of adherent mammalian cells on a substrate. PloS one 13:e0192265

Bauer R (2021) The Future of Cryopreservation: Computational Approaches and Automatisation. Biostasis the annual biostasis conference, Zurich

Bojic S et al (2021) Winter is coming: the future of cryopreservation. BMC Biol 19, 56

Braslavsky I, Lipson SG (1998) Electrofreezing effect and nucleation of ice crystals in free growth experiments. Appl Phys Lett 72:264-266

Chow R et al (2003) The sonocrystallization of ice in sucrose solutions: primary and secondary nucleation. Ultrasonics 41:595-604

Costanzo JP, Lee RE Jr (1995) Supercooling and Ice nucleation in vertebrate ectotherms. In Lee RE Jr. et al (eds) Biological ice nucleation and its applications, APS Press, St. Paul MN, pp 221-238

Costanzo JP et al (1995) Survival mechanisms of vertebrate ectotherms at subfreezing temperatures: Applications in cryomedicine. FASEB J 9:351-358

Dalvi-Isfahan M et al (2017) Review on the control of ice nucleation by ultrasound waves, electric and magnetic fields. J Food Eng 195:222-234

De Wolf A, de Wolf C (2013) Human cryopreservation research at advanced neural biosciences. In Sames KH (ed) Applied Human Cryobiology, vol. 1, ibidem, Stuttgart, pp 45-59

Fahy GM (2021) Principles of vitrification as a method of cryopreservation in reproductive biology and medicine. In Donnez J, Kim SS (eds) Section 2—Reproductive Biology and Cryobiology. Cambridge University Press, pp 49-66

Fuller BJ et al (2019) Ch. 22.6.1 corneas. In BJ et al (eds) Life in the Frozen State, (2004) 1st ed. CRC Press, Boca Raton

Han X et al (2007) Effects of nanoparticles on the nucleation and devitrification temperatures of polyol cryoprotectant solutions. Microfluid Nanofluidics 4:357

Hochi S et al (2001) Effects of cooling and warming rates during vitrification on fertilization of in vitro-matured bovine oocytes. Cryobiology 42:69-73

Hübinger J (2013) Bedeutung der Vitrifikation für die Kryokonservierung und Kryofixierung. Dissertation, Technische Universität Dortmund

Jones AG (2002) Crystallization process systems. Butterworth-Heinemann, Oxford

Jang TH et al (2017) Cryopreservation and its clinical applications. Integr Med Res 6:12-18

Karlsson JEM, Toner M (1996) Long-term storage of tissues by cryopreservation: critical issues. Biomaterials 17:243-256

Karlsson JO et al (1993) Nucleation and growth of ice crystals inside cultured hepatocytes during freezing in the presence of dimethyl sulfoxide. Biophys J 65:2524-2553

Karthika S et al (2016) A review of classical and nonclassical nucleation theories. Cryst Growth Des 16:6663-6681

Kashchiev D (2000) Nucleation: basic theory with applications. Butterworth-Heinemann, Oxford

König O (1997) Equipment for controlling nucleation and tailoring the size of solution-grown single crystals. J Appl Crystallogr 30:507-509

Kuleshova LL et al (1999) Sugars exert a major influence on the vitrification properties of ethylene glycol-based solutions and have low toxicity to embryos and oocytes. Cryobiology 38:119-130

Lindinger B et al (2007) Ice crystallization induced by optical breakdown. Phys Rev Lett 99:045701

Lundheim R (2002) Physiological and ecological significance of biological ice nucleators. Philos Trans R Soc Lond B Biol Sci 357:937-943

Luyet B (1937) The vitrification of organic colloids and of protoplasm. Biodynamica 1:1-14

Lynch ME, Diller KR (1981) Analysis of kinetics of cell freezing with cryoprophylactic additives. Trans ASME 81-WA/HAT-53

Margaritis A, Bassi AS (1991) Principles and biotechnological applications of bacterial ice nucleation. Crit Rev Biotechnol 11:277-295

Mazur P (1963) Kinetic of water loss from cells at subzero temperatures and the likelihood of intracellular freezing. J Gen Physiol 47:347-369

Mazur P (1984) Freezing of living cells: mechanisms and implications. Amer J Physiol 247 (3 Pt 1): C 125-142

Mazur P et al (1972) A two-factor hypothesis of freezing injury. Exper Cell Res 1:345-355

McFarlane DR (1987) Physical aspects of vitrification in aqueous solutions. Cryobiology 23:181-195

Meryman HT (1970) The exceeding of a minimum tolerable cell volume in hypertonic suspension as a cause of freezing injury. In Wolstenholme GEW, O'Connor M (eds) The frozen cell. Ciba Foundation Symposium. Churchill, London, pp 51-64

Morris GJ, Acton E (2013) Controlled ice nucleation in cryopreservation—a review. Cryobiology 66:85-92

Morris GJ et al (2006) Cryopreservation of murine embryos, human spermatozoa and embryonic stem cells using a liquid nitrogen-free, controlled rate freezer. Reprod Biomed Online 13:421-426

Muldrew K, McGann LE (1994) The osmotic rupture hypothesis of intracellular freezing injury. Biophys J 66:532-541

Muldrew K et al (2004) The water to ice transition: implications for living cells. In Fuller BJ et al (eds) Life in the Frozen State. Boca Raton, CRC Press

Paynter SJ et al (1999) Temperature dependence of Kedem-Katchalsky membrane transport coefficients for mature mouse oocytes in the presence of ethylene glycol. Cryobiology 39:169-176

Petersen et al (2003) Controlled nucleation in freezing biological material. Cryobiology 47:256

Petersen A et al (2006) A new approach for freezing of aqueous solutions under active control of the nucleation temperature. Cryobiology 53:248-257

Petrenko VF, Whitworth RW (1999) Physics of ice. Oxford University Press (OUP)

Plumb RC (1989) Antique windowpanes and the flow of supercooled liquids. J Chem Educ 66:994-996

Pradzynski CC et al (2012) A fully size-resolved perspective on the crystallization of water clusters. Science 337:1529-1532

Saragusty J (2015) Directional freezing for large volume cryopreservation. Methods Mol Biol 1257:381-397

Spindler R et al (2011) Dimethyl sulfoxide and ethylene glycol promote membrane phase change during cryopreservation. Cryo Letters 32:148-157

Sputtek A (1996) Cryopreservation of blood cells. In: Müller-Eckhard C (ed) Transfusion medicine, basics, therapy, methodology, 2nd ed. Springer, Berlin, pp 125-135

Steif PS et al (2007) The effect of temperature gradients on stress development during cryopreservation via vitrification. Cell Preserv Technol 5:104-115

Steponkus PL et al (1983) Destabilization of the plasma membrane of isolated plant protoplasts during a freeze-thaw cycle: the influence of cold acclimation. Cryobiology 20:448-465

Vali G (1995) Principles of ice nucleation. In: Lee RE Jr, Warren GJ (eds) Biological Ice Nucleation and Its Applications. APS Press, St. Paul Minnesota, pp 1-28

Vali G (1996) Ice nucleation-a review. In Kulmala M, Wagner PE (eds) Nucleation and Atmospheric Aerosols. Pergamon, Amsterdam, pp 271-279

Wesley-Smith J et al (2015) Why is intracellular ice lethal? A microscopical study showing evidence of programmed cell death in cryo-exposed embryonic axes of recalcitrant seeds of Acer saccharinum. Ann Bot 115:991-1000

Wolkers WF, Oldenhof H (2021) Principles Underlying Cryopreservation and Freeze-Drying of Cells and Tissues. Methods Mol Biol 2180:3-25

Wolkers WF et al (2007) Effects of freezing on membranes and proteins in LNCaP prostate tumor cells. Biochim Biophys Acta 1768:728-736

Wowk B (2007) How cryoprotectants work. Cryonics 3rd Quart 2007 (www.Alcor.org)

Wowk B, Fahy GM (2007) Ice nucleation and growth in concentrated vitrification solutions. Cryobiology 55:330 (Abstr. 21)

Yu G et al (2017) Characterizing intracellular ice formation of lymphoblasts using low-temperature Raman spectroscopy. Biophys J 112:2653-2663

Zachariassen KE, Kristiansen E (2000) Ice nucleation and antinucleation in Nature. Cryobiology 41:257-279

Zanotto ED (1998) Do cathedral glasses flow? Amer J Physics 66:392-395

6 Tools for cryonics: Cryoprotectants — A big step forward but not yet the complete solution to our problems

Summary

Cryoprotective agents were one of the first ways to intervene decisively in the crystallization process of water, to keep ice crystals in check by lowering the freezing point, and thus to mitigate the greatest difficulty in deep-cooling biological cells and tissues. While they can only prevent crystals before reaching cryogenic temperature ranges and to a limited extent at such temperatures, they do allow advances in organ preservation, the most difficult near-term goal of cryonics. Unfortunately, they are not harmless. Most importantly, they can cause damage where they must be used in high concentrations. Apart from that, certain cryoprotective agents can cause specific damage to different parts of the tissues. Among other things, a rough overview of the very different effective substances involved in cryonics will be given here.

The melting temperature and the temperatures at which water freezes due to various crystallization nuclei are all lowered by cryoprotectants. With sufficient cryoprotectant, the temperature of the freezing point could coincide with the glass transition temperature (see for vitrification) (Fahy et al. 1984). This would be ideal for cryonics if concentrations and cooling rates in the tissue could be controlled accordingly.

However, a somewhat less ideal practical possibility arises. This is because if the freezing point is lowered, a tissue sample can be kept at lower temperatures but above the freezing point in supercooled fluid, which is free of ice crystals. Supercooling strategies are discussed clearly by Taylor et al. (2019).

Cryoprotective agents reduce damage by replacing water and, on the other hand, by diluting dissolved substances to avoid harmful concentrations. They reduce ice formation by forming hydrogen bonds with water molecules. They thus prevent water from binding to ice crystals. Scientifically speaking, antifreeze agents can interrupt ice formation by colligative interference with hydrogen bonds (Best 2015).

Cryoprotective agents conveniently lower both the freezing point and the melting point of aqueous solutions. The melting point of an aqueous solution decreases rapidly with increasing concentration of the cryoprotectant. In a (deceased) patient perfused with glycerol, ice melting occurs near-60 °C. A concentration of cryoprotectant that pushes the melting point to -30°C lowers the freezing point to the glass transition temperature.

Unfortunately, cryoprotectants do not completely prevent crystal formation (see also Best B vitrification in cryonics. https://www.benbest.com/cryonics/vitrify.html).

However, they can lead to higher survival rates when cooled slowly. For example, increasing the glycerol concentration from 0.4M to 1.25M increases cell survival (Sputtek 1996). So, they actually bring progress to cryonics.

Cryoprotectants must be added before a deep cooling and diluted again during thawing. This causes changes in osmotic pressure inside and/or outside the cells. Thus, one should not proceed too quickly.

A high water-binding capacity is an indication of a substance's effectiveness as a cryoprotectant.

> Background information
> Chemical groups found in various substances that form strong hydrogen bonds with water such as: Hydroxyl, amide and sulfoxide groups make up a high water content.
> The physical and chemical antifreeze properties of hydrocarbon derivatives of diols, glycerol, amides, and sulfoxides have long been extensively tested by researchers (Pichugin 1993).
> The water-attracting properties are probably effective as a prerequisite for the antifreeze promoting so-called colligative qualities.

Hydrophobic properties, on the other hand, have a negative effect, e.g. on membranes, fat-protein compounds and newly formed proteins.

Water repellency generally increases with the length of groups containing many carbon atoms, but such long chains pass the cell membrane with greater difficulty (Connor, Achwood-Smith 1973; Doebbler 1966; Jeyendran, Graham 1977; Lovelock 1953; Pichugin, Novikov 1989).

6.1 Small but effective

Cryoprotectants that are able to penetrate cells are called penetrating cryoprotective agents (CPAs). They can considerably promote cell survival. As early as 1949, Polge et al. found that glycerol promoted the resuscitation of spermatozoa exposed to low temperatures by removing water from the cells and replacing the water in the cells. This prevented ice formation.

Different antifreeze substances now pass through the cell membrane at different speeds. They even change the rate at which water penetrates the cell membrane (Gilmore et al. 1995; Vian, Higgins 2014). At medium concentrations of cryoprotectants such as DMSO or ethylene glycol, pores are formed in membranes, which increase the hydraulic permeability of the membranes (Spindler et al. 2011; Kharasch, Thyagarajan 1983).

Rapid migration across the cell membrane is a desirable property of antifreeze substances. It depends on the active chemical groups in the molecules.

Small molecules that are readily accessible to cells and hydrophilic include, for example, substances with alcoholic groups such as polyols or diols (ethanediol, propanediol, butanediol, glycerol).

Such cryoprotectants, which are able to penetrate into the cells, can prevent excessive water loss from the cell (as occurs with slow cooling) by retaining water in the cell (the water loss itself, however, depends on many factors, including temperature).

Thus, these cryoprotective solutions lower the so-called equilibrium temperature, which depends on their concentration. Thereby water loss from the cell contents only occurs at lower temperatures when ice crystals form outside the cells and attract water to themselves. Such cryoprotectants thus counteract the osmotic equilibrium disturbance during freezing (Pegg 2007).

At low temperatures, the hydraulic permeability of membranes is reduced because it decreases with temperature. This also causes water to remain in the cell. The risk due to cell shrinkage is reduced. Cryoprotectants that penetrate the cells naturally lower the freezing point in the interior of the cell as well. Its temperature can be lowered in insulated cells to the glass transition temperature of the liquid to achieve successful deep cooling (Sputtek 1996).

6.2 What makes a cryoprotectant?

The water-binding capacity and cell passability have already been discussed.

Important for antifreeze are: the colligative properties. The colligative effect of substances is the effect which depends on the number of particles, i.e. not on the weight or volume. The effect of these particles consists, for example, in the fact that particles of a certain substance practically stand in the way of particles of water during ice formation.

Good membrane permeability and low deleteriousness, lowering of the freezing point, and water-binding capacity as well as viscosity in solution play a major role in the effectiveness of a cryoprotectant (see in Karlsson, Toner 1996; Pegg 2007, 2015; see in Sputtek 1996).

> Background information
> Alcohols with attached chemical methoxyl groups (amides and sulfoxides) are less harmful and enter cells better (dimethylformamide, dimethylacetamide, dimethyl sulfoxide (DMSO) (Fahy et al.1987; Forsyth, MacFarlane 1990; Jeyendran, Graham 1977; Karow 1969; Wowk; Wowk 1999).

The cell-permeable cryoprotectants are supposed to draw water into the cell. However, as mentioned, they are often small molecules that bind little water. It is therefore difficult to find those with sufficient water-binding capacity. The damaging effect of antifreeze substances usually increases with the concentration of the molecules. However, the quantity by weight of a substance with small molecules must contain more molecules than the same quantity of a substance with large molecules. Therefore, substances with small molecules can more easily cause damage (Lovelock 1953; Meryman 1971).

However, cryoprotective solutions pass through membranes at different rates even without regard of their size, for example, ethylene glycol enters cells more quickly than glycerol.

> Background information
> For dihydric alcohols (diols), it has been shown that their ability to suppress crystallization depends on the strength of hydrogen bonding. Stronger bonding inhibits the movement of water molecules and thus the formation of crystallization nuclei and crystal growth. However, if the bond becomes too strong, hydrates form, which then limit the ability to vitrify (see below) (Forsyth, McFarlane 1990).

Cryonics can take advantage of the fact that small amounts of cryoprotectant can be sufficient during slow cooling, since the concentration of substances in the cell also acts as antifreeze in the process. Namely, if cryoprotectant has already penetrated the cell and the cell now suffers water loss due to ice formation outside the cell during slow cooling, the solution in the cell becomes thicker and less cryoprotectant is needed. For example, DMSO, which is harmful but well cell penetrable, can be provided in harmless concentrations. DMSO passes through the cell membrane more easily than glycerol, but is more harmful. A combination of both is particularly refined.

According to what has been said, there are two ways to protect cells during slow cooling:

a) so much water is removed from them that the substances in the cell become highly concentrated and thus practically act as antifreeze to lower the freezing point.
b) Cryoprotectants penetrate the cell and if water is now withdrawn, this promotes the effect of the penetrated antifreeze.

6.3 Large cryoprotectant molecules also have advantages

Cryoprotectants with large molecules are e.g. mostly polymers e.g. polyvinylpyrrolidone (PVP), polyethylene oxide (PEO), polyethylene glycol (PEG), trehalose, dextranes, chemically modified gelatin, hydroxyethyl starch (HES), albumin protein (Sputtek 1996) or proteoglycans, the latter being present everywhere in the tissue. These substances do not penetrate cells and are therefore easier to remove. HES, proteoglycans and albumin protein are tolerated by the body and do not need to be removed.

At first glance, it would seem that such large molecules could do less because they are held outside the cells.

Unfortunately, it is not known in full detail how such large molecules nevertheless act as antifreeze for cells. There are several possible explanations for this.

Some cryoprotectants such as trehalose or hydroxyethyl starch, which do not penetrate the cell membrane, are theorized to strengthen protein substances through a preferential exclusion effect and processes of thermal physics, or to stabilize cell membranes through electrical action. They have been found to prevent damage during deep cooling best in combination with cryoprotective agents such as DMSO.

Such substances draw water out of the cell, as does the formation of ice crystals (see above). The ingenious thing is that they prevent ice crystals at the same time. With these cryoprotective substances, therefore, a lower concentration of antifreeze is needed inside the cells.

Then the water removal can cause vitrification (see below) to take place within the cell. They further promote vitrification by increasing the glass transition temperature (see Wolkers, Oldenhof 2021). Vitrification outside the cell with a sugar prevents cell membranes from coming into contact and fusing (Crowe et al. 1998; - 2003).

In general, such substances are less harmful than those which penetrate into the cells, for example, because fewer large molecules fit into the same amount of water. That is why they are often used in mixed antifreeze solutions.

Some non-cell penetrating substances additionally reduce chilling injury.

It is therefore worthwhile for cryonics to get to know antifreeze substances in detail.

In 2022 deep eutectic systems containing e.g. trehalose, glucose, sorbitol, proline, glycerol and others have been shown to protect cell cultures during cryopreservation (Jesus et al. 2022; Bryant et al. 2022)

6.4 Molecule gathering up: individual cryoprotective agents and their effect

Overall, we have a wide range of antifreeze substances with different effects. Unfortunately, you cannot use them all at once, but there are quite effective cryoprotective mixtures. We will discuss individual cryoprotectants here. The description of individual cryoprotectants shows how diverse their applications are in controlling tissue cooling and protecting cells. It should not be withheld from the interested reader.

> Background information
> 2,3-Butanediol is a strong antifreeze. Unfortunately, it contains structural types (isomers) that promote ice formation and is expensive in purified form (Sutton 1992). However, like sucrose (see below) and other substances, it can greatly reduce ice crystal formation (Wowk et al. 2018). For cooling and heating, there are different critical rates in tissues for various cryoprotectants for which crystallization during cooling and heating have been tested e. g. with butanediol on kidney tissue (Peyridieu et al. 1996).
> Amides are weak cryoprotectants compared to polyalcohols (polyols).

Dimethyl sulfoxide (DMSO) is one of the most commonly used cryoprotectants. Currently, it is the preferentially used cryoprotectant for stem cell storage (Bakken 2006). It was first described as an antifreeze by Lovelock and Bishop (1959). DMSO, as mentioned, has the advantage of increased membrane permeability for many cell types compared to glycerol (Baust et al. 2009; Ock, Rho 2011; Polge et al. 1949). It is also present in the cryoprotective solutions used to perfuse the bodies of the deceased. DMSO can have damaging effects (Harriger et al. 1997; Karow et al. 1967). It probably alters cell membranes by removing water (Westh 2004) and reduces ice crystal formation (Klbik et al. 2022). Unlike the multiple alcohols glycerol, ethylene glycol, propylene glycol, and others, the harmfulness of DMSO can be reduced by mixing it with other cryoprotectants. Heat is generated during the mixing process. The degree of heat thereby indicates the extent of the reduction in harmful effect (Fahy et al. 1987). However, DMSO-free cryoprotectant solutions or those with reduced DMSO concentrations have also been developed (Svalgard et al. 2020; Weng, Beauchesne 2020).

Sweet-and-sour antifreeze: the antifreeze effect of sugars

Single sugars (monosaccharides), double sugars (disaccharides) such as sucrose and trehalose, triple sugars (e.g., raffinose), and multiple sugars (sugar polymers or polysaccharides) such as starch have been used as antifreeze agents (Kuleshova et al.1999).

> Background information
> Glycerol is so widely used that it only needs to be mentioned here. Glycerol harms cell membranes the least among cryoprotectants with small molecules. More critical is the effect of dimethysulfoxide (DMSO) or ethylene glycol during slow cooling (Hess et al. 2004).
> Dextrose (glucose) is found in animals as a cryoprotectant. A glucose that does not participate in the metabolic flux, i.e. is not consumed (3-o-methyl-n-glucose =3OMG) is particularly suitable as a cryoprotectant and has been successfully used together with polyethylene glycol (PEG) for liver antifreeze protection (Berendsen et al. 2014).
> Disaccharides such as sucrose and trehalose with molecules consisting of two individual sugars cannot cross the cell membrane.
> Sucrose consists of fructose-glucose units and trehalose of glucose-glucose. Like butanediol (see above) and other substances, it can reduce ice formation when added to the newly developed DP6 antifreeze solution.

Sucrose and trehalose best protect proteins and membranes against chilling injury, ice formation and dehydration. In cell membranes, the two sugars bind to the head groups of phosphorus-containing lipids (phospholipid molecules) and stabilize them.

Trehalose has amazing properties. Slow freezing studies showed that trehalose can alter the response between cells and ice, resulting in improved cell survival (Arakawa, Timasheff 1983; Hubel et al. 2007; Mantri et al. 2015). It is a harmless sugar with cryoprotective activity. It binds more strongly than water to protein and lipid components of cell membranes at the electrically active ends of this membrane molecules. However, trehalose does not cause deformation (denaturation) of proteins in the process, as do other cryoprotectants (Rudolp et al. 1986). Its vitrification properties can be controlled (Weng, Elliot 2015). Human reproduction could also be positively influenced by trehalose. Microinjection of trehalose into human oocytes and combination with DMSO in mouse oocytes improve cryopreservation (Eroglu et al. 2002; -2009). Therefore, attempts are also being made to introduce trehalose into cells by various means (Gao et al. 2022; Wolkers, Oldenhof 2021). Trehalose can also act as a radical scavenger to protect large molecules of tissues against free radicals (Benaroudj et al. 2001), a property that is desirable in dying tissues. Thus, one could kill two birds with one stone (antifreeze and detoxification). The large water space (hydrated volume) found around trehalose removes a lot of water from ice formation (Jain, Roy 2009; Sei et al. 2002).

Hydroxyethyl starch (HES) also does not penetrate cells. It is a harmless plasma expander — used for cryopreservation of red blood cells — and does not need to be removed or diluted, like glycerol which needs to be concentrated very high to be effective (Scott et al. Space 2005). However, glycerol is only harmful in such high concentrations.

Background information
Flavone-like polymer sugar compounds from the glycoside group, allow the freezing temperature to be lowered by up to 9°C (Kasuga et al. 2010).

Polysaccharides, like almost all sugars, increase the glass transition temperature (Kuleshova et al. 1999). It is therefore possible to cool more slowly in order to reach the glass transition temperature.

Sugars and proteins are also an energy reserve. In nature, does this energy help in resuscitation after waking up before an animal is vigorous enough to seek food? Biodegradable polysaccharides with antifreeze activity can also be suitable for food preservation. Guerreiro et al. (2018) have developed such a polysaccharide. However, starch or glycosaminoglycans are also biodegradable.

Our body is already a framework of antifreeze and fibers

Acid sugars: the antifreeze effect of proteoglycans

Background information
Proteoglycans are a group of the many proteins conjugated with polymer sugars and they are normal building blocks of our tissues. They are mainly located outside and around the cells, but are also found inside the cells and even in the cell nuclei. Proteoglycans also have many important functions apart from forming structure in our tissues. Their action as cryoprotectants is also well known.

Proteoglycans are giant biological molecules consisting of protein chains to which side chains of acidic polysaccharides are attached (see Sames 1994).

These sugar side chains—polymers containing multiple sugars called mucopolysaccharides or glycosaminoglycans—carry negatively charged side chains on at least every second individual sugar molecule, which makes them acidic. These chains are chondroitin and chondroitin sulfate, dermatan sulfate, keratan sulfate, heparin and heparan sulfate, and hyaluronic acid. The latter is used as a cosmetic due to its strong water binding properties, which plumps the skin and reduces wrinkles. Its effect is not permanent. So, in the sense of the providers a lot of "hyaluron ointment" is consumed.

Heparin is known as an anticoagulant. Other glycosaminoglycans also inhibit blood clotting but more moderately (with lower side effects).

Background information
It is slowly becoming clear today what role proteoglycans and similar substances play. They are found everywhere in the space

between the cells in the normal structure of our tissues, even in bones and teeth.
Together with collagen and elastin fibers surrounding the cells, they make up the bulging elasticity of our body. Glycosaminoglycans tend to occupy as large a water space as possible (in some cases 300 times their own volume). This hinders the formation of ice.
Thus, these substances help build the entire framework of our body. There is antifreeze everywhere. In addition, there are the numerous sugars and proteins of other classes. The fact that a human body is nevertheless not frost-proof need not be mentioned, but perhaps the body's own substances could be used in frost protection. Perhaps they are just not present in high enough concentrations to provide complete frost protection, but it doesn't have to be that simple.
It would be interesting to use a mixture of long chains, double sugars and single sugars of the glycosaminoglycans, which penetrate the cells to varying degrees according to their size. The protein of the proteoglycans could also be tested for its antifreeze effect. When the cells are alive, the single sugars in the cell can be linked to form chains, which the cell releases to the outside.
By using chemicals that prevent their elimination from the cell, these long chains can also be accumulated in the cells. Do they thus protect the cell during cryopreservation? This can only be attempted as long as the cells are alive.
Proteoglycans have been successfully tested as cryoprotectants e.g. chondroitin sulfate in the storage of corneal tissue at 4°C (Fuller BJ et al. 2004; Lindstrom 1990).
They change the type of ice crystal formation in solutions. Normally, many crystals form longitudinally oriented patterns that grow through the tissue without regard to the cells. With glycosaminoglycans, they become cross-linked or randomly oriented patterns (Allenspach, Kraemer 1989).
The number of surviving bovine embryos (day 8) after freezing and thawing increased from 11 to 75% (!) by addition of hyaluronic acid to synthetic tubal fluid, and the changes seen at high magnification as well as the loss of specific genetic and other properties of the cells (degeneration or de-differentiation) were reduced (Stojkovic et al. 2002).

We have found excellent preservation of the tissue without any cryoprotectant by light microscopy on cartilage. This concerns the basic substance (matrix) outside the cells, which consists of proteoglycans up to 40% of the dry weight.

One could assume that cartilage is the frost protection master among the tissues. However, its weak point is its easily-damaged cells. The preservation of cells is not ideal when cooled without additional cryoprotectant. Actually, glycosaminoglycans should extract water from the cells, but they are in equilibrium with the space inside the cells from the formation of the tissues. They may well limit water loss from the cells under normal conditions by maintaining this equilibrium. Tissues with extremely high levels of proteoglycans, such as cartilage, may be naturally protected during temperature degradation. One might conclude that about 30% proteoglycans allow slow freezing. In most tissues, therefore, proteoglycans would still have to be added if they were to be used as cryoprotectants. This is speculation at the moment.

The high proteoglycan content makes cryopreservation of cartilage particularly interesting. It is therefore already discussed here.

Background information
Cartilage retained its mechanical properties when cryopreserved with DMSO (Kiefer et al. 1989). This may be explained by the cryoprotective properties of proteoglycans in the matrix and additional protection of the cells by DMSO. The survival of cartilage after thawing and transplantation has been confirmed several times especially when 10% DMSO is used. This cryoprotectant however, is not completely harmless to the cells (Abazari et al. 2013; Jackson et al. 1992; Kawabe, Yoshinao 1990; Kiefer et al. 1989; Kushibe et al. 2001; Schachar et al. 1992; Sharma et al. 2007) whereby total cell loss may occur (Arnoczky et al. 1988; -1992; Stevenson et al. 1989). However, cartilage cells also survive appropriate freezing methods while maintaining high viability, preserving their ability to produce proteoglycans (Almqvist 2001; Schachar et al. 1989). They can survive to as much as 90% (Tomford 1984; Tomford et al. 1985).

The antifreeze effect of the body's own antifreeze agents and their exploitation for cryopreservation remains to be investigated. Since there are creatures that survive the lowest temperatures with the aid of their own bodies (see below), this could be promising.

6.5 Side effects of cryoprotectants

Unfortunately, cryoprotectants also have effects that can damage biological tissues. This is a difficult situation for cryonics, since it are precisely cryoprotective solutions that are used to prevent cells and tissues from being destroyed by ice crystals during cooling. Thus, the harmfulness of cryoprotectants is a major challenge for cryopreservation (Best 2015). In particular, it threatens to jeopardize the great progress made by vitrification and much work has been done to identify such side effects.

However, science does not have to give up. There are ways to mitigate side effects. They can be dampened or compensated for, for example, in such a way that cell cultures survive the deep-freezing process. However, one has to know quite a bit about the behavior of the cryoprotectants and, on the other hand, about the reaction of biological tissues to these substances.

The harmful effect of cryoprotectants decreases as the temperature drops. On the other hand, their viscosity in solution increases. So here, too, we are walking a fine line. If perfusion is run at somewhat higher temperatures in order to pump more antifreeze through the vessels, one risks an increased damaging effect of the antifreeze. At lower temperatures, they become more viscous and pass through the blood vessels more poorly (see also Khandoga 2003).

An open general question is what damage one has to accept when freezing human bodies and whether one learns to repair it in the future, i.e. what can be tolerated.

Furthermore, the harmfulness of cryoprotectants is different for various cell types and tissues (often sperm and oocytes or embryos are used for testing). For example, ethylene glycol is not harmful to bovine embryos (Gilmore et al. 1997). Glycerol is the least harmful to kidney slices compared to other substances. Also, in different animal species, the harmfulness of different cryoprotectants differs even to the same tissues.

For human cells of the inner wall of blood vessels, even at 4°C, DMSO and ethylene glycol — the cryoprotectants of the Cryonics

Institute solution — are more harmful than propylene glycol or 2,3-butanediol (Wusteman MC et al. 2002), but have other advantages.

The treatment of tissues and organs, which contain many different cells and tissue components, is difficult.

> Background information
> Harmful reactions that result from general mechanisms caused by properties common to different antifreeze agents (such as water binding or change of osmotic pressure with increasing concentration, or forming hydrogen bonds with proteins as described by Arakawa (2007) form the non-specific toxicity.
> Harmful properties typical for individual cryoprotectants, on the other hand, form the specific toxicity.
> The temperature dependence of the effects of cryoprotectants is important. Prescriptions according to which cryoprotectant is added at different times and in different amounts can affect the viability of tissues after thawing. In addition, because of the different properties of cell types and tissues, it is critical that the correct cocktail of cryoprotectants is applied in a manner that is appropriate to the properties of the sample (Fahy, Wowk 2015).

Because so many aspects have to be considered simultaneously, recent studies have also used methods with a high processing capacity, such as gene expression analyses, to determine the harmfulness of cryoprotectants. It is important to note here that harmful effects depend primarily on the level of concentrations (Cordeiro 2015; Warner et al. 2021).

The harmful behavior of cryoprotectants becomes a problem especially at higher temperatures or with slow cooling and at high concentrations. Reducing the concentration or shortening the exposure time (e.g. by rapid cooling) can help. Cooling during oxygen starvation reduces harmful reactions during a subsequent perfusion (Khandoga 2003).

The side effects may be amplified in large objects such as organs due to uneven distribution. Therefore, it is good to have methods that make the latter visible (Corral et al. 2021).

6.6 Critical concentration changes of cryoprotectants

The addition and dilution of cryoprotectant leads to strong concentration changes (osmotic stress).

Dilution of cryoprotectants allows the cell to absorb water and swell. When cryoprotectant is added, water is first drawn out of the cells. When diluted, the solution passes through tissue, and water permeates into the cell, which swells. Substances that do not pass through the cell membrane, such as sucrose, can act as so-called osmotic buffers during dilution and retain water outside the cells (Armitage 1986 and see below).

Swelling during dilution of cryoprotective solutions is — as mentioned — more dangerous for the cell than the shrinkage when concentrated cryoprotectants are added. Cells lining the blood vessels (endothelial cells), which are important for open passage of blood vessels, are, like other cells, more damaged by swelling than by shrinkage.

Since water passes through the cell membrane faster than almost all solutes (Armitage 1986), when cryoprotectant molecules are added and enter the cell environment, water first escapes from the cell. Once the cryoprotectant molecules have then entered the cell, water is again drawn in and this then causes cells to swell. Therefore, they must be added and removed slowly. They accumulate in front of the cell, so to speak, and withdraw water from it, causing it to shrink.

Time is needed for both the accumulation and dilution of the cryoprotectants. This also means that the cryoprotectants or the diluted solution should circulate with the bloodstream for a longer period of time during cryonics of human bodies. In this way, an extensive equalization can take place. The permeability of the cell membrane plays a major role in this process. However, a recent paper demonstrated that permeation is very fast as long as tissue is perfused by liquid (Bleisinger 2020).

In the case of the intricately composed tissues, other components besides the cell membranes play a role, e.g., the permeability of the ground substance (matrix) between the cells (mainly fibers and proteoglycans), which acts like a space-sized sieve, is important. Of course, the more or less long distances that a cryoprotectant has to travel from the blood vessel to the cells also have an effect. Furthermore, chemical substances of the tissue that react with the cryoprotectant have an influence.

This is another reason why imaging techniques that attempt to visualize antifreeze in tissue are so interesting (Risco 2021; see in Wolkers, Oldenhof 2021).

In cryonics today, the highly concentrated cryoprotectant solution is left in the body. However, this creates a problem in the future when tissues will be thawed again and water enters the cells. Fortunately, more cells survive high osmotic pressure at low temperatures than at room temperature (Zawlodska, Takamatsu 2005).

Considering the limited resistance of different cells to the high concentrations of cryoprotectants, these must be added and removed in the form of a concentration gradient. This will slowly adjust the sample to the cryoprotectants to reduce exposure to the changes in osmotic pressure (Fahy, Wowk 2015). One should give cryoprotectants at low concentrations at first, which one then increases (a procedure called "ramping"). This reduces the damaging effect. If cryoprotectant is added quickly, more cells die than if it is added slowly (Muldrew, Mc Gann 1990; -1994). An addition in steps is less favorable than a steady increase (Bourne et al. 1994). During addition, imbalances in the concentration of cryoprotectants between different tissue regions can affect the success of cryopreservation (Fahy et al. 2009; Fahy et al. 1984; Rall 1987).

Addition and dilution of cryoprotectants can therefore have multiple lethal effects on cells in a purely osmotic manner, i.e. through changes in the concentration of substances. This illustrates the problems we are getting into. It is a miracle that cryobiology works at all. In cryonics, one must remember that many cells are already inactive (dormant) or dead after cardiac arrest, possibly without changing further once the body has been cooled. However, they may passively participate in changes (Burg et al. 1997; -2007; Copp. et al. 2005; Gao et al. 1995; Liu, Foote 1998; Mullen et al. 2004; Pollock et al. 1986).

The membranes of cells respond differently to various cryoprotectants. The cell membrane of human spermatozoa, for example, is four times more permeable to ethylene glycol compared with glycerol and this permeability is less temperature-dependent than that for glycerol (Gilmore et al. 1997).

In addition to the membranes of their cells, the capillaries of the bloodstream have other partially permeable membranes. These can allow other substances to pass through than the cell membranes. The so-called blood-brain barrier, which consists of membranes of the capillaries and glial cells, has a somewhat different permeability. The permeability for water is reduced here, as is that for active molecules (electrolytes). In this way, the nerve cells of the brain, which are electrically charged by ion distribution, receive chemical protection.

Osmotic pressure within the capillaries is caused mainly by proteins that are too large to leave the circulation through the membranes (oncotic pressure). This pressure is about 28mm mercury column at normal blood pressure.

Both aqueous cell swelling (cellular edema) and aqueous tissue swelling, in which water also accumulates outside the cells (tissue edema), can impede flow in the capillaries.

It was also found, e. g. with liver cells, that the water-binding in cell cultures is completely different from that in cells located in the tissue. Thus, the concentration equilibrium with a cryoprotectant is probably also changed and one can transfer results on cell cultures to body cells only with caution (own observation and see Karlsson, Toner 1996).

The causes of the harmfulness of cryoprotectants are still sparsely researched, despite some knowledge. It is desirable to find explanations based on the structure of the individual cryoprotective molecule for all substances (Fahy et al. 1987).

Of course, it is difficult to determine the harmfulness of cryoprotectants unless the living beings are rewarmed and -animated. That is why it is difficult to make statements about cryonics in humans. So far, conclusions can only be drawn from experiments on cell cultures or isolated organs.

6.6.1 What is actually damaged and how?

Relationships have been found between the harmfulness of cryoprotectants and the stability of protein components in cell membranes in contact with cryoprotectants (Therefore, the harmfulness of cryoprotectants was thought to result from changes in proteins (protein denaturation hypothesis). However, this probably does not apply to temperatures and concentrations during vitrification and thus only to the initial course of the cryonics method (Arakawa et al. 1990; Fahy et al. 1990; Gilmore et al. 1995; -1997; Ivanov 2001).

> Background information
> Certain cryoprotectants damage cell membranes and the so-called skeleton of the cell (cytoskeleton) by their specific toxicity. They inhibit the transmission of information from the program on the DNA into the cell and its organelles. They change the electrical charge of the membrane on the respiratory organelles and directly damage DNA and proteins. Cell division can be hindered and even cell suicide (apoptosis) can be induced. When antifreeze solution is added quickly, the inner wall cells of the vessels shrink so that the cell connections (junctions) break, which weakens the protective function of the cell layer.

Gregory Fahy and Brian Wowk have found and patented a measure of the damaging effect of antifreeze solutions, the qv^*.

According to their theory, the damage depends on the electrically activatable (polar) chemical groups in antifreeze solutions, measured at concentrations required for vitrification. The formula is $q = Mw/Mpg$.

Mw is the weight of water in one liter of the antifreeze solution expressed in moles. Mpg is the weight of the active groups also in one liter of the solution (in moles). The mysterious qv^* is q at the concentration needed to vitrify 5-10 ml of antifreeze at a cooling rate of 10°C/minute.

We hope this gives us a real way to determine the harmful effects of all cryoprotective solutions with one formula based on their chemical values, without the need for biological experiments. In any case, individual facts can be explained by qv. For example, cryoprotective solutions with a high q have fewer electrically active groups. They bind more strongly to water, which increases their damaging effect. As is well known, water is needed by many molecules of living tissue for their activity.

The best properties of cryoprotective solutions include low harmfulness and low viscosity (Fahy et al. 2004; see also Best 2015).

6.7 Increasing of the cryoprotectant concentration protection or death?
Reduction of damaging effects
Reduction of harmful effects

Cryoprotectant solutions are often very highly concentrated to promote uptake into the cells. The high concentration increases the damaging effect.

However, there are other ways to offset the harmful effects besides cooling.

Fortunately, mixtures of cryoprotective substances are often less damaging than individual substances. In this case, the specific harmfulness of each substance can be diluted, so to speak, by the other substances, while the desired effect is enhanced by all substances — an old trick of drug manufacturers. A mixture, in other words, reduces the harmfulness of the individual substances because each causes a different kind of harm, but they all produce the same benefit. A reduced concentration of the individual components logically allows a higher total concentration of a mixed solution without increasing the harmfulness. Above all, it is favorable to mix small molecules that go into the cell and large molecules that do not, and also to add those with strong vitrification effects.

Some cryoprotectants reduce the harmfulness of others. For example, the damaging effect of formamide is reduced by both glycerol (Warner et al. 2021) and DMSO.

> Background information
> It has been found that using ethylene glycol instead of propylene glycol reduces the harmfulness of a solution because ethylene glycol forms weaker hydrogen bonds than propylene glycol. This

means that large molecular components of cells and tissues lose less water (Best 2015; Fahy 2010; Fahy et al. 2004).

The M22 mixture used by the cryonics organization Alcor for cryopreservation of deceased (suspension) is an advanced legally protected solution developed by the research organization 21Century Medicine.

> Background information
> The solution contains the methoxylated multiple alcohol 3-methoxy-1,2 propanediol. The addition of a methoxyl group changes the behavior of an alcohol. Thus, 3-methoxy-1,2-propanediol gains increased ability to pass into cells and a decreased viscosity. Reduced viscosity is important because it provides the ability to the cryoprotectant of perfusing the tissues. This represents a major advantage for cryonics. In addition, propanediol lowers the critical rates of heating and cooling of solutions i.e., there is more time for cryoprotectants to soak through the tissues (Fahy et al. 2004; Fahy 2006; Wowk et al. 1999).

Most of the cryoprotectants in the M22 antifreeze solution go into the cell with the exception of PVP, K12 and ice blockers.

VM1, developed by Yuri Pichugin on behalf of CI, is a simple solution (CI-VM-1). It contains 35% ethylene glycol and 35% dimethyl sulfoxide. This combination reduces the harmfulness. Ethylene glycol has a beneficial effect on DMSO (Gautam et al.2008). VM-1 is unfortunately more harmful than M22 but also has advantages.

> Background information
> In 2014, Brockbank tested on cartilage and blood vessels modified solutions developed by Fahy with propylene glycol and formamide, DMSO, and highly concentrated VS55. He obtained high vitality with good preservation of the tissue matrix.

The newer DP6 solution takes into account a number of the mentioned possibilities and shows particularly favorable characteristics (Wowk 2018).

Overall, it appears that ways to reduce the damaging effects of cryoprotectants can be developed. In doing so, it is beneficial to know the effect of cryoprotectants on the cell during cryopreservation (Elliot et al. 2017).

Literature

Abazari A et al (2013) Cryopreservation of articular cartilage. Cryobiology 66:201-209

Allenspach AL, Kraemer TG (1989) Ice crystal patterns in artificial gels of extracellular matrix macromolecules after quick-freezing and freeze-substitution. Cryobiology 26:170-179

Almqvist KF et al (2001) Biological freezing of human articular chondrocytes. Osteoarthritis Cartilage 9:341-350

Arakawa T, Timasheff SN (1983) Preferential interactions of proteins with solvent components in aqueous amino acid solutions. Arch Biochem Biophys 224:169-177

Arakawa T et al (1990) The basis for toxicity of certain cryoprotectants: A hypothesis. Cryobiology 27:401-415

Arakawa T et al (2007) Protein precipitation and denaturation by dimethyl sulfoxide. Biophys Chem 131:62-70

Armitage WJ (1986) Osmotic stress as a factor in the detrimental effect of glycerol on human platelets. Cryobiology 23:116-125

Arnoczky SP (1992) Cellular repopulation of deep-frozen meniscal autografts: An experimental study in the dog. Arthroscopy 8:428-436

Arnoczky SP et al (1988) The effect of cryopreservation on canine menisci: a biochemical, morphologic, and biomechanical evaluation. J Orthop Res 6:1-12

Bakken AM (2006) Cryopreserving human peripheral blood progenitor cells. Curr Stem Cell Res Ther 1:47-54

Baust JG et al (2009) Cryopreservation: an emerging paradigm shift. Organogenesis 5:90-99

Benaroudj N et al (2001) Trehalose accumulation during cellular stress protects cells and cellular proteins from damage by oxygen radicals. J Biol Chem 276:24261-24267

Berendsen TA et al (2014) Supercooling enables long-term transplant survival following 4 days of liver preservation. Nat Med 20:790-793

Best BP (2015) Cryoprotectant toxicity: facts, issues, and questions. Rejuvenation Res 18:422-436

Bleisinger N et al (2020) Me2SO perfusion time for whole-organ cryopreservation can be shortened: results of micro-computed tomography monitoring during Me2SO perfusion of rat hearts. PLoS ONE 15 e0238519, https://doi.org/10.1371/journal.pone.0238519

Bourne WM et al (1994) Human corneal endothelial tolerance to glycerol, dimethylsulfoxide, 1,2-propanediol, and 2,3-butanediol. Cryobiology 31:1-9

Brockbank KGM (2014) Tissue vitrification C-20 (conference abstract). Cryobiology 69:50

Bryant SJ (2022) Deep eutectic solvents as cryoprotective agents for mammalian cells. J Mater Chem B 10:4546-4560

Burg MB et al (1997) Regulation of gene expression by hypertonicity. Annu Rev Physiol 5:437-455

Burg MB et al (2007) Cellular response to hyperosmotic stresses. Physiol Rev 87:1441-1474

Connor KW, Achwood-Smith ML (1973) Ineffectiveness of dimethyl sulfones as a cryoprotective agent. Cryobiology 10:87-89

Copp J et al (2005) Hypertonic shock inhibits growth factor receptor signaling, induces caspase-3 activation, and causes reversible fragmentation of the mitochondrial network. Amer J Physiol Cell Physiol 288:C403-C451

Cordeiro RM et al (2015) Insights on cryoprotectant toxicity from gene expression profiling of endothelial cells exposed to ethylene glycol. Cryobiology 71:405-412

Corral A et al (2021) Use of X-ray computed tomography for monitoring tissue permeation processes. In Wolkers WF, Oldenhof H (eds) Cryopreservation and Freeze-Drying Protocols. Methods Mol Biol 2180:317-330

Crowe et al (1998) The role of vitrification in anhydrobiosis. Annu Rev Physiol 60:73-103

Crowe JH et al (2003) Stabilization of membranes in human platelets freeze-dried with trehalose. Chem Phys Lipids 122:41-52

Doebbler GF (1966) Cryoprotective compounds: review and discussion of structure and function. Cryobiology 3:2-11

Elliot GD et al (2017) Cryoprotectants: a review of the actions and applications of cryoprotective solutes that modulate cell recovery from ultra-low temperatures. Cryobiology 76:74-91

Eroglu A (2002) Beneficial effect of microinjected trehalose on the cryosurvival of human oocytes. Fertil Steril 77:152-158

Eroglu A et al (2009) Successful cryopreservation of mouse oocytes by using low concentrations of trehalose and dimethylsulfoxide. Biol Reprod 80:70-78

Fahy GM (2006) Cryopreservation of complex systems: the missing link in the regenerative medicine supply chain. Rejuvenation Res 9:279-291

Fahy GM (2010) Cryoprotectant toxicity neutralization. Cryobiology 60 (3 Suppl) p 45-53

Fahy GM, Wowk B (2015) Principles of Cryopreservation by Vitrification. In Wolkers WF, Oldenhof H (eds) Cryopreservation and freeze-drying protocols. Methods Mol Biol 1257:21-82

Fahy GM et al (1984) Vitrification as an approach to cryopreservation. Cryobiology 21:407-426

Fahy GM et al (1987) Some emerging principles underlying the physical properties, biological actions, and utility of vitrification solutions. Cryobiology 24:196-213

Fahy GM et al (1990) Cryoprotectant Toxicity and Cryoprotectant Toxicity Reduction: In Search of Molecular Mechanisms. Cryobiology 27:247-268

Fahy GM et al (2004) Improved vitrification solution based on the predictability of vitrification solution toxicity. Cryobiology 42:22-35

Fahy GM et al (2009) Physical and Biological Aspects of Renal Vitrification. Organogenesis 5:167-175

Forsyth M, MacFarlane DR (1990) A study of hydrogen bonding in concentrated diol/water solutions by proton NMR correlations with glass formation. J Physical Chem 9:6889-6893

Fuller BJ et al (2019) Ch. 22.6.1 corneas. In Fuller BJ et al (eds) Life in the Frozen State, 1st ed (2004), CRC Press, Boca Raton.

Gao DY et al (1995) Prevention of osmotic injury to human spermatozoa during addition and removal of glycerol. Hum Reprod 10:1109-1122

Gao S et al (2022) Cryopreservation of human erythrocytes through high intracellular trehalose with membrane stabilization of maltotriose-grafted ε-poly(L-lysine.). J Mater Chem B 10:4452-4462, doi: 10.1039/d2tb00445c.

Gautam SK et al (2008) Effect of type of cryoprotectant on morphology and developmental competence of in vitro-matured buffalo (Bubalus bubalis) oocytes subjected to slow freezing or vitrification. Reprod Fertil Dev 20:490-496

Gilmore JA et al (1995) Effect of cryoprotectant solutes on water permeability of human spermatozoa. Biol Reproduct 53:985-995

Gilmore JA et al (1997) Determination of optimal cryoprotectants and procedures for their addition and removal from human spermatozoa. Hum Reprod 12:112-118

Guerreiro BM et al (2018) A novel polysaccharide-based approach for cryopreservation. Cryobiology 85:125

Harriger MD et al (1997) Reduced engraftment and wound closure of cryopreserved cultured skin substitutes grafted to athymic mice. Cryobiology 35:132-142

Hess R et al (2004) Ethylene glycol: an estimate of tolerable levels of exposure based on a review of animal and human data. Arch Toxicol 78:671-680

Hubel A et al (2007) Cell partitioning during the directional solidification of trehalose solutions. Cryobiology 55:182-188

Ivanov IT (2001) Rapid method for comparing the cytotoxicity of organic solvents and their ability to destabilize proteins of the erythrocyte membrane. Pharmacy 56:808-809

Jackson DW et al (1992) Meniscal transplantation using fresh and cryopreserved allografts. An experimental study in goat. Am J Sports Med 20:644-656

Jain NK, Roy I (2009) Effect of trehalose on protein structure. Protein Sci 18:24-36

Jesus AR et al (2022) Use of natural deep eutectic systems as new cryoprotectant agents in the vitrification of mammalian cells. Sci Rep 12: 8095

Jeyendran RS, Graham EF (1977) Cryoprotective compounds for bull spermatozoa. Cryobiology 14:703 Conference abstract

Karlsson JEM, Toner M (1996) Long-term storage of tissues by cryopreservation: critical issues. Biomaterials 1:243-256

Karow AM (1969) Cryoprotectants — a new class of drugs. J Pharm Pharmacol 21:209-213

Karow AM Jr et al (1967) Toxicity of high dimethyl sulfoxide concentrations in rat heart freezing. Cryobiology 3:404-468

Kasuga J et al (2010) Analysis of supercooling-facilitating (anti-ice nucleation) activity of flavonol glycosides. Cryobiology 60:240-243

Kawabe N, Yoshinao M (1990) Cryopreservation of cartilage. Int Orthop 14:231-235

Kharash N, Thyagarajan BS (1983) Structural basis for biological activities of dimethyl sulfoxide. Ann N Y Acad Sci 411:391-402

Khandoga A et al (2003) Impact of intraischemic temperature on oxidative stress during hepatic reperfusion. Free Radic Biol Med 35:901-909

Kiefer GN et al (1989) The effect of cryopreservation on the biomechanical behavior of bovine articular cartilage. J Orthop Res 7:494-501

Klbik I (2022) On crystallization of water confined in liposomes and cryoprotective action of DMSO. RSC Adv 12:2300-2309

Kuleshova LL et al (1999) Sugars exert a major influence on the vitrification properties of ethylene glycol-based solutions and have low toxicity to embryos and oocytes. Cryobiology 38:119-130

Kushibe K et al (2001) Tracheal allograft maintaining cartilage viability with long-term cryopreserved allografts. Ann Thorac Surg 71:1666-1669

Lindstrom RL (1990) Advances in corneal preservation. Trans Am Ophthalmol Soc 88:555-648

Liu Z, Foote RH (1998) Bull sperm motility and membrane integrity in media varying in osmolality. J Dairy Sci 81:1868-1873.

Lovelock JE (1953) The mechanism of protective action of glycerol against hemolysis by freezing and thawing. Biochem Biophys Acta 11:28-36

Lovelock JE, Bishop MW (1959) Prevention of freezing damage to living cells by dimethyl sulphoxide. Nature 183:1394-1395

Mantri S et al (2015) Cryoprotective effect of disaccharides on cord blood stem cells with minimal use of DMSO. Indian J Hematol Blood Transfus 31:206-212.

Meryman HT (1971) Cryoprotective agents. Cryobiology 8:173-183

Muldrew K, McGann LE (1990) Mechanisms of intracellular ice formation. Biophys J 57:525-532

Muldrew K, McGann LE (1994) The osmotic rupture hypothesis of intracellular freezing injury. Biophys J 66:532-541

Mullen SF et al (2004) The effect of osmotic stress on the metaphase II spindle of human oocytes, and the relevance to cryopreservation. Hum Reprod 19:1148-1154

Ock SA, Rho GJ (2011) Effect of dimethyl sulfoxide (DMSO) on cryopreservation of porcine mesenchymal stem cells (pMSCs). Cell Transplant 20:1231-1239

Pegg DE (2007) Principles of cryopreservation. Methods Mol Biol 368:39-57

Pegg DE (2015) Principles of cryopreservation. In Wolkers WF, Oldenhof H (eds) Cryopreservation and freeze-drying protocols. Methods Mol. Biol. 1257:3-19

Peyridieu JF et al (1996) Critical cooling and warming rates to avoid ice crystallization in small pieces of mammalian organs permeated with cryoprotective agents. Cryobiology 33:436-446

Pichugin YI, Novikov AN (1989) The dependence of cytotoxicity and cryoprotective efficiency of diols on their structure and physico-chemical properties. In Kharkov (ed) Cryopreservation of Cells and tissues, pp 15-28

Pichugin Y (1993) Results and perspectives in searching of new endocellular cryoprotectants. Problems of Cryobiology (Kharkov, Ukraine) 2:3-9

Polge C et al (1949) Revival of spermatozoa after vitrification and dehydration at low temperatures. Nature 164:666

Pollock GA et al (1986) An isolated perfused rat mesentery model for direct observation of the vasculature during cryopreservation. Cryobiology 23:500-511

Rall WF (1987) Factors affecting the survival of mouse embryos cryopreserved by vitrification. Cryobiology 24:387-402

Risco R (2021) Session 3—Experimental Evidence of Return to Life with High intensity Focused Ultrasound. Biostasis the annual biostasis conference, Zurich

Rudolp AS et al (1986) Effects of three stabilizing agents--proline, betaine, and trehalose--on membrane phospholipids. Arch Biochem Biophys 245:134-143

Sames K (1994) The role of proteoglycans and glycosaminoglycans in aging. Interdiscip Top Gerontol Geriatr vol. 28. Karger, Basel

Schachar N et al (1989) Cryopreserved articular chondrocytes grow in culture, maintain cartilage phenotype, and synthesize matrix components. J Orthopaedic Research 7:344-351

Schachar N et al (1992) Metabolic and biochemical status of articular cartilage following cryopreservation and transplantation: a rabbit model. J Orthop Res 10:603-609

Sharma R et al (2007) A novel method to measure cryoprotectant permeation into intact articular cartilage. Cryobiology 54:196-203

Scott KL et al (2005) Biopreservation of red blood cells: past, present, and future. Transfus Med Rev 19:127-142 Review

Sei T et al (2002) Growth rate and morphology of ice crystals growing in a solution of trehalose and water. J Cryst Growth 240:218-229

Spindler R et al (2011) Dimethyl sulfoxide and ethylene glycol promote membrane phase change during cryopreservation. Cryo Letters 32:148-157

Sputtek A (1996) Cryopreservation of blood cells. In Müller-Eckhard C (ed) Transfusion Medicine, Fundamentals, Therapy, Methodology, 2nd ed. Springer, Berlin, pp 125-135

Stevenson S et al (1989) The fate of articular cartilage after transplantation of fresh and cryopreserved tissue-antigen-matched and mismatched osteochondral allografts in dogs. J Bone Joint Surg Am 71:1297-1307

Stojkovic M et al (2002) Effects of high concentrations of hyaluronan in culture medium on development and survival rates of fresh and frozen-thawed bovine embryos produced in vitro. Reproduction 124:141-153

Sutton RL (1992) Critical cooling rates for aqueous cryoprotectants in the presence of sugars and polysaccharides. Cryobiology 29:585-598

Svalgaard JD et al (2020) Cryopreservation of adipose-derived stromal/stem cells using 1-2% Me2SO (DMSO) in combination with pentaisomaltose: An effective and less toxic alternative to comparable freezing media. Cryobiology 96:207-213

Taylor MJ et al (2019) New approaches to cryopreservation of cells, tissues, and organs. Transfus Med Hemother 46:197-215

Tomford WW (1984) Studies on cryopreservation of articular cartilage chondrocytes. J Bone Joint Surg Am 66:253-259

Tomford WW et al (1985) Experimental freeze-preservation of chondrocytes. Clin Orthop 197:11-14

Vian AM, Higgins AZ (2014) Membrane permeability of the human granulocyte to water, dimethyl sulfoxide, glycerol, propylene glycol and ethylene glycol. Cryobiology 68:35-42

Warner RS et al (2021) Rapid quantification of multi-cryoprotectant toxicity using an automated liquid handling (2015)ng method. Cryobiology 98:219-232

Weng L, Beauchesne PR (2020) Dimethyl sulfoxide-free cryopreservation for cell therapy: a review. Cryobiology 94:9-17

Weng L, Elliot GD (2015) Different glass transition behaviors of trehalose mixed with Na_2HPO_4 or NaH_2PO_4: Evidence for its molecular origin. Pharm Res 32:2217-2228

Westh P (2004) Preferential interaction of dimethyl sulfoxide and phosphatidyl choline membranes. Biochim Biophys Acta 1664:217-223

Wolkers WF, Oldenhof H (2021) Principles Underlying Cryopreservation and Freeze-Drying of Cells and Tissues. Methods Mol Biol 2180:3-25

Wowk B et al (1999) Effects of solute methoxylation on glass-forming ability and stability of vitrification solutions. Cryobiology 39:215-227

Wowk B et al (2018) Vitrification tendency and stability of DP6-based vitrification solutions for complex tissue cryopreservation. Cryobiology 82:70-77

Wusteman MC et al (2002) Vitrification media: toxicity, permeability, and dielectric properties. Cryobiology 44:24-37

Zawlodzka S, Takamatsu H (2005) Osmotic injury of PC-3 cells by hypertonic NaCl solutions at temperatures above 0 degrees C. Cryobiology 50:58-70.

7 Other methods for protecting the cells and preventing ice crystals

Summary

Cryoprotectants only partially solve the problems of cryopreservation. But there are other ways to make cryopreservation possible. For example, cryoprotectants can also help with vitrification. In addition, ice blockers are known. Actions on cell membranes can serve to protect the membranes or optimize their permeability to cryoprotective substances. Special reactions of molecules lead to the formation of clathrates, which exclude ice crystals but demand a lot of space. We already know a lot of different effects of cryoprotectants as well as of different cooling rates, concentrations and reactions of cells and by dimensions different sizes of objects. In order to develop optimally combined approaches to such large amounts of data, information programs are being sought. Obviously, there is still room for problem solutions. Again and again, it is discussed whether a chemical fixation could not protect the tissues faster and more effectively than the unstable deep-cooling.

7.1 Protection of vital membranes

The cell membrane is what makes life possible. It excludes physical and chemical effects from the environment that can distort the direction of metabolic processes.

Since the cell membrane is a particular target of freezing damage, cryoprotectants are favorable which, in addition to preventing ice formation, also protect the cell membranes (e.g. protection of the cell membrane by ethylene glycol). So-called 3-block polymers such as P188 are effective even at concentrations of thousandths of a mole (Lee 2002).

There are also agents to stabilize the cell membrane against chilling injury and ice formation.

Background information
Such are: proline, betaine, sarcosine, glycerol, DMS0. Trehalose and sucrose. They partially prevent membrane fusion. Proline, betaine and sarcosine also stabilize the membrane's phosphorus-containing fatty substances (phospholipids) by water-repellent (hydrophobic) reactions (Anchordoguy et al.1987).

Membrane protection is also served by relatives of cortisone such as dexamethasone.

7.2 Chemical fixation?

Fixatives preserve those molecules with which they react, e. g. glutaraldehyde, the proteins by reaction with amino acids. Lipids and carbohydrates do not take part in this. They can even be lost.

Chemical fixation could rapidly arrest the changes that result from the cessation of blood flow after death certification. For a long time, however, it was argued that such methods promoted ice formation in the cells in experiments.

Now methods exist that prevent ice formation and a new investigation is interesting.

In experiments, the flow through fixed tissues was unimpeded if the stop of blood flow before fixation was not too long. After cooling to the temperature of liquid nitrogen, no ice crystal formation occurred even after 2 weeks in the cold. The particularity was that tissue swelling was absent. However, there was also a strong water loss of the fixed brain. Even a substance to open the so-called blood-brain barrier, so that water can flow over this barrier (i.e. from the blood into the tissue and the cells) did not change anything.

However, if a delay occurred between death certification and fixation, the well-known formation of ice crystals occurred.

With one hour delay, severe flow obstructions and ice crystals were already occurring.

Samples for the electron microscope were still taken hours after death and storage at room temperature.

The consequence of the results is again that cooling (or fixation) should be performed as early as possible and blood should be replaced directly on site with organ protection solution (washout) in order to reduce the obstruction of the flow and the formation of ice crystals after blood exchange and cooling (de Wolf, de Wolf 2013).

What is serious in the case of fixation or similar chemical reactions is the strong change in molecular structures, such as protein structures and in water binding due to chemical reactions. How these could be reversed upon resuscitation remains an open question. E. g., the author has previously shown that imaging of proteoglycans (and thus images of connective tissue) with standard contrast and precipitation methods of electron microscopy represent collapse artifacts. They are called matrix granules and are usually accepted as normal tissue structure. However, it was shown that they can be avoided by appropriate methods (Sames 1990; -1994 p.37). Galhuber et al (2021) found a method to preserve ultrastructure during cryopreservation. Fine connections are often difficult to represent with standard methods. Artifacts probably arise from precipitation of, for example, protein molecules. It is difficult to tell what is artifact and what corresponds to normal structure, and it is not clear whether artifacts could be "deconvoluted", e.g., to create a real image of the connectome and finally allow tissue water uptake (rehydration) and reanimation.

7.3 Methods to cope with huge amounts of data

Unfortunately, numerous factors influence the deep-cooling of biological objects. We have already mentioned a number of them.

> Background information
> It is not only the temperature or cooling rate. It is also crystallization nuclei, properties of cryoprotectants, the mutual influence of cryoprotectants on each other, the concentration of cryoprotectants, properties of liquids and solutes, osmotic pressure, properties of cell membranes and the liquid in the cells (cytosol), the structure of the tissues especially those of the blood circulation, the permeability of biological structures and the substances which build up cells and tissues and—quite decisively—the size of the objects (see also Wolkers, Oldenhof 2021).

Finding the ideal combination of all possibilities and even improving them is hardly possible by experiments and mathematically too extensive. So far, one starts from the best results and tests their further improvement in experiments, which has already yielded astonishing progress (Gautam et al. 2008).

In this situation, the use of computer programs is more and more envisaged, which today can already standardize biological tests with large amounts of data (see e.g. Luechtenfeld et al. 2018). However, it is sometimes difficult to put the mentioned properties into numbers or to consider the mutual influences together e. g. in studies called -omics.

The existing programs are mostly concerned with biochemical properties or with the work of the brain or synthetic biology research. Similar methods have also been useful in cryopreservation. Pioneering work in this field was done by P. Mazur (1963). He found that there were different ideal cooling rates for different, individual cell types. This led to mathematical formulas for the relationship between ideal cooling rates and cell properties. More recently, such models have been used to establish time sequences for the addition and removal of cryoprotectants. Similarly, the relationship of cooling velocity and temperature has been determined for immersion in liquid nitrogen, for example (Benson et al. 2012; Kashuba et al. 2014; Mazur 1990).

Simulations of molecular processes have also been used to track the antifreeze effects of different concentrations of cryoprotectants and temperatures. This can help understand the influence of cryoprotectants at the molecular level. A computer model has also been used to study the relationship of temperature stress to volume (Bojic et al. 2021; Feldschuh et al. 2005; Solanki et al. 2017; Valojerdi et al. 2009; Weng et al. 2011).

In complexly built tissues, for example, it would be important to find lowest common denominators for the reactions of different cells or building units.

However, one should not expect a quick breakthrough. So far, only individual problems have been tackled by computer models in extensive work. However, the latter are already a valuable supplement to the experiments, as can be seen from the examples.

7.4 How cryoprotectants help with glass formation (vitrification)

It is possible to promote vitrification, because a high viscosity can be achieved even before cooling by a strong concentration of cryoprotectants (see below). (Wowk et al. 2000; Fahy et al. 2004). Normally, one chooses a concentration of 30-50% (maximum 9M).

Since the cells do not tolerate such high concentrations well, the gradual addition of solutions is even more important during vitrification than during slow cooling.

Viscosity as an important criterion for the effect of cryoprotective solutions depends not only on the concentration. Specific properties of cryoprotectant molecules also play a role.

A decreasing range of viscosity is: glycerol > propylene glycol > ethylene glycol > DMSO

The glass transition temperature ranges from -80 to- 130°C. It was previously shown that vitrification of small biological samples (e.g. human sperm) can be achieved without cryoprotectant when using an extremely high cooling rate (Nawroth et al. 2002).

The general approach—especially for organ cryopreservation—is to use lower concentrations first to allow time for uptake into the cells and then, when the temperature has become lower, use higher concentrations to further remove water from the cells (Guerreiro et al. 2016).

In addition to lowering the freezing point of the solution, cryoprotectants prevent ice formation by increasing the glass transition temperature. Through these effects, they also lower the necessary cooling rate and allow the sample to form a glass before ice formation can occur (Courbiere et al. 2006; Fahy, Wowk 2015). This also provides a way to reduce the problems of large samples to some degree, as a decreased cooling rate always creates more time for temperature equilibration even in larger samples.

The biggest problem with vitrification then, however, is that very high concentrations of cryoprotectant are harmful. Adding cryoprotectant only at a low temperature can of course reduce the harmful effect (Best 2015; Fahy et al.1990; Wowk 2010).

Different cryoprotectants have different readiness to form glass, e.g. in 45% solution. This decreases in the following order of substances:

propylene glycol > DMSO > DMF > 1,4-butanediol > ethylene glycol > glycerol > 1,3-propanediol

It has been shown on kidney sections that those antifreeze agents that bind most strongly to the hydrogen of water also vitrify best (Fahy et al. 2004).

But one must not rejoice too soon, because the same cryoprotective agents also react strongly with the hydrogen of protein compounds and can change their shape, which is important for functioning. They could perhaps also extract water from tissues and thus damage tissues. Annoyingly, therefore, just the most harmless cryoprotectant vitrifies only weakly.

The following series lists substances according to their harmful effects. It decreases from left to right:

Formamide> Propylene glycol> DMSO (Dimethyl sulfoxide)> Ethylene glycol> Glycerin.

Interestingly, this is also approximately the order of glass-forming ability (Baudot 2000). Formamide solution, however, cannot vitrify.

In sections from the kidney, the better the solutions form glass, the more harmful they prove to be.

Vitrification solutions can indeed promote the vitrification of larger samples. However, these are far from samples the size of human organs. Samples with large volume relative to surface area cool delayed internally at least when not well perfused (Hopkins et al 2012; Kilbride et al 2016).

M22 was used to vitrify a rabbit brain without ice formation (Lemler et al. 2004).

VM1 is a stronger and less expensive cryoprotectant than M22. It also vitrifies better. Unfortunately, it is more harmful.

Similar to the rapid decrease in temperature, an increase in pressure favors viscosity, promotes glass formation and hinders crystal formation. Therefore, increasing the pressure and simultaneously lowering the temperature can be used in combination during vitrification. Originally, there was no other method. The vitrification solution VS4 used in the past vitrified at a pressure of 100 atmospheres. Such pressures are technically difficult to realize for large objects.

A good example here of solving problems that initially seem insoluble was the development of a new cryoprotective solution. For vitrification in the laboratory, mixtures of DMSO, formamide and propanediol (VS41A) were dosed to 55 percent by weight in aqueous salt solution. At a pressure of 1 atmosphere, they acted as VS4 would at 100 atmospheres. Thus, the pressure could be normalized (Kheirabdi, Fahy 2000; Mehl 1993).

The suitability of proteoglycans for vitrification has not yet been tested.

Ice crystals can unfortunately also be formed in a mixture used to achieve vitrification. This is found, for example, when a cryonics patient is insufficiently perfused.

With M22, nuclei of ice crystallization can form between -100- and -135°C. At -140°C, ice formation in M22 ceases. The crystallization nuclei consist of local nests of crystallized water molecules, which can then promote the growth of more ice crystals in the course of a warming up. They themselves are harmless, but indicate a mobility of molecules that could be harmful.

It is not only cryoprotectant that supports glass formation. Carrier solutions can reduce the amount of cryoprotectant required for vitrification. This dilutes the harmful effect of cryoprotectants, so to speak. In fact, water vitrifies more easily when it contains salt or cryoprotectant. Salt solutions have their highest glass transition temperature at their eutectic concentration. However, this is too high for cryobiological applications.

Solutions containing different salts vitrify at different temperatures, depending on chemical properties of the salts (Angell, Sare 1970).

How the glass transition temperature is influenced by the nature of the cryoprotectant present and that of other solutes can be easily observed. For example, it is 10-20°C higher for propylene glycol than for glycerol. Concentrated cryoprotective solutions with high viscosity also reduce the formation of crystallization nuclei and thus inhibit crystallization. Glass formation in such high viscous solutions can occur even at lower cooling rates, because increasing viscosity of liquid is part of the vitrification.

According to all that has been said so far, crystal growth can thus be avoided during cooling by two measures:

a) due to a high speed of cooling
b) due to high concentration or high viscosity of cryoprotectants.

Suitable concentrations are about 6-8 moles per liter. As a rule of thumb, the concentration that is practically useful to achieve vitrification may be 10 times that for a freeze-thaw protocol (see above) with slow freezing (see Karlson and Toner 1996; Kheirabadi, Fahy 2000).

The glass transition temperature (T_g) increases with the concentration of cryoprotectant, which means that it is reached earlier during cooling, while the temperature for ice crystal formation decreases.

Since crystal growth decreases with viscosity, it must also decrease with decreasing temperature, because solutions become more viscous in the process. When the glass transition temperature is reached, there is hardly any movement of the molecules and hardly any crystal formation. Ice crystals can therefore be formed between the melting point and the glass transition temperature. We can shorten the period for this by rapid cooling (Uhlmann 1972).

Glycerol would not crystallize at 68% of the volume fraction of a solution in water at any temperature below 0°C. However, glycerol concentrations above 55% are too viscous and toxic to flow through the blood circuit. American cryobiologist Brian Wowk showed that a combination of 58.4% (8 molar) glycerol and 1% of the ice blocker X100 in a solution can vitrify the contents of a 2 L bottle, so this would purely by mass be also possible with a brain. By the way, at high concentrations glycerol does not enter the cells very much and pulls water out of the cells, which leads to dehydration of the cells. You cannot vitrify with glycerol because, although enough of it enters the cells to perform partial vitrification when the highest concentration is used, about 20% of the water forms ice (see also Best B Vitrification in cryonics. https://www.benbest.com/cryonics/vitrify.html).

A vitrification solution usually contains cryoprotectant a physiological (isotonic) salt solution and one or more solutes with large molecules e.g. polyethylene glycol with addition of PVP, Ficoll or dextran. In this case, the required concentration can be a sum of the concentrations of the different molecules with similar effects (Shaw et al. 1997). It is important that the glassy nature is maintained during thawing at the melting point.

Ice crystal formation is suppressed by common solutions when cooling down to e.g. -125°C, where the glass transition temperature is reached at which no more crystal formation occurs. Fast cooling can play a less important role in such solutions as the solutions improve. One example is the new DP6 solution.

An additional role is played by all possible additives from tissues, blood or tissue culture fluids.

All in all, the above information shows that one can attack in many places to avoid harmful side effects but often with little room for maneuver. Precise knowledge and persistent experimentation can lead to success even in difficult terrain. There is no need to give up one's optimism in cryonics endeavor, and we still have other possibilities to discuss (see Fahy et al. 2004b).

7.4.1 Bound water

The removal of water generally does not immediately lead to harmful dehydration of the cell, but if the bound water is removed, such damage occurs. Water can be included in crystals of many inorganic and organic molecules (hydrate forms, from Greek hydor = water). Protein crystals, for example, usually contain 50% water. Large molecules (macromolecules) start to react chemically with each other when they lose the protective shell of (bound) water. Most water is unbound in living tissues. Bound water is so viscous that it vitrifies easily without forming crystals. At least 10% of cellular water (the bound portion) is therefore incapable of freezing (i.e. forming ice crystals).

> Background information
> In solutions of biopolymers such as starch, gluten, collagen, albumin protein, etc. of high molecular weight, the glass transition temperature for the bound water is around -10°C
> Aqueous solutions of cryoprotective agents can also freeze. Here, too, the glass transition temperature for the bound water is higher than the glass transition temperature for the total water. The portion of a sample concentrated at maximum by freezing naturally has a high viscosity (Bai et al 2001; Brake, Fenema 1999; Goff 1995).

When ice crystals bind water from the free solution, the unfrozen portion that is present next to it gains a higher solute concentration due to water removal. This also causes a higher glass transition temperature in this region (Izutsu et al. 2009; Wowk 2010).

If a sample contains ice crystals in addition to residual unfrozen solution, the unfrozen portion also consists of bound water (see also Best B Physical parameters of cooling in cryonics. https://www.benbest.com/cryonics/cooling.html; Schreuders et al 1996; Sun 1999).

7.4.2 Ice crystals are not tasty

Crystallization also plays a role in one of the most popular delicacies that civilization offers us, namely ice cream, crystallization plays a role especially during thawing. Ice cream is naturally accompanied by a liquid phase, which creates a softer mass. Additionally, ice blockers made of winter wheat grass are used to prevent crystallization in ice cream (very young wheat, Regand 2006). For ice cream, the non-frozen water represents 35% of the weight (Fennema 1996).

If a patient devitrifies during thawing (see below), ice will be found in some regions, in others the highly enriched unfrozen solution, which—taken by itself—can vitrify already at -110°C (instead of e.g. at -123°C for M22 at Alcor or CI-VM1 at Cryonics Institute). The storage temperature should in any case be below the glass transition temperature, i.e. from about -138°C (see Sputtek 1996).

7.5 Another advance: No more ice crystals? The ice blockers

Ice blockers are a special type of non-membrane-permeable cryoprotectants because they bind directly to ice crystals and other crystallization nuclei (heteronuclei). They represent the synthetic equivalent of antifreeze proteins, which occur naturally (Wowk 2007). The non-membrane permeating ice blockers such as polyvinyl alcohol or polyglycerol are thus a cheaper and more effective alternative to animal antifreeze proteins. The addition of ice blockers in small amounts prevents crystallization by exogenous crystallization nuclei. Substances such as the disaccharide trehalose or ice blockers thus inhibit crystals outside the cells where more crystallization nuclei are present (Fahy, Wowk 2015). Trehalose, which does not go into cells can prevent crystal growth along the A axes of ice crystals better than sucrose and inhibits cell suicide, decreases ice crystal size, and promotes cell viability (Sei et al. 2002; Shinde et al. 2019; Solocinski et al. 2017).

Various animals are known to produce substances, including proteins, that allow them to survive low temperatures by preventing ice formation. Amino acids such as threonine and serine in the protein chain molecules form hydrogen bonds with ice (Zachariassen, Kristiansen 2000).

Arctic fish use proteins to maintain the freezing temperature of their bodies at -2.2°C or below, i.e., below the temperature of their sea water (-1.9°C). The antifreeze substances can be glycoproteins, i.e., protein molecules with side chains of sugars, which include the proteoglycans mentioned above, and we also carry such substances in our bodies. Ice blockers cannot be used in cryonics at cryogenic temperatures, but it is being investigated whether they could complement other antifreeze agents (Guerreira et al. 2016; Tas et al. 2021).

Using ice blockers in cryonics could bring further progress. However, obtaining such substances from animal material is expensive. Therefore, plant materials have also been investigated (Kawahara 2008), but synthetic materials have been preferred. For example, polyvinyl alcohol inhibits crystal growth and polyglycerol inhibits crystal nucleation. They are effective even in small amounts.

While other antifreeze agents allow ice crystal growth, ice blockers act specifically against the formation of ice crystallization nuclei in the form of ice crystals without which ice crystal growth in pure water will not start.

They thus inhibit the progression of crystallization, whereas normal cryoprotectants retard crystallization mainly during cooling, so that it only occurs at lower temperatures.

In the upper temperature range below 0°C, ice blockers offer the possibility of reducing the formation of crystallization nuclei outside the cells and in the bloodstream. As a result, up to 55% of cryoprotectant can be saved during vitrification This means a reduction in harmfulness. Ice blockers can both bind and render harmless crystallization nuclei and stop crystal growth along various axes.

In vitrification, therefore, attempts are being made to reduce harmful effects by adding ice blockers (e.g. polyvinyl alcohol, see Wowk et al. 2000). Polyvinyl alcohol proved to be particularly suitable for use in vitrification (Naitana et al. 1997).

The strongest formation of crystallization nuclei occurs just above the glass transition temperature (- 85 – -120°C). The strongest growth of ice crystals, on the other hand, occurs just below the melting point (-40 – -80°C).

Ice crystals are therefore much easier to avoid during cooling than during rewarming (devitrification), because during heating, crystallization nuclei are first absorbed and then you enter the zone of crystal growth.

Before the use of ice blockers, it was believed that only high-speed heating e.g. radio frequency heating could avoid crystal formation during heating. The use of ice blockers is a real progress.

However, they do not totally prevent crystal growth during heating.

Ice blockers can act in three ways of which two have already been mentioned a) by binding crystallization nuclei b) by blocking the a-axes c) by blocking the c-axes of ice crystals (Zachariassen, Kristiansen 2000). Indeed, ice crystals can grow around 6 axes lying in the same plane, the a-axes, and one axis perpendicular to it, the c-axis. At higher temperatures, ice crystal growth occurs along the a-axes, which makes up the hexagonal image of snowflakes. The c-axis growth leads to needle-shaped crystals, which can be particularly harmful (Davies, Hew 1990).

We touch here a complicated field of physics, which interests us only as far as it can be applied to the cooling of biological entities.

In the meantime, there have been many special studies on antifreeze proteins as used by living organisms in nature (see below and Crevel 2002).

> Background information
> When added to glycerol solutions, for example, polymers such as polyvinyl alcohol prevent ice formation and crystallization even at low doses. It seems that destruction of cells and tissues by crystallization can be avoided and freeze fractures are also reduced (Wowk et al. 2000). 21st Century Medicine markets a mixture of 20% vinyl acetate and 80% vinyl alcohol as Supercool X-1000. It can lower the concentration of glycerol needed for glycerol-containing vitrification mixtures. Polyglycerol from 21st Century Medicine binds and inactivates proteins that can serve as condensation nuclei in ice formation (Wowk, Fahy 2002).

Since ice blockers do not pass the blood-brain barrier, they can only be found in the vascular system of the brain, which represents only 4% of the brain. However, ice blockers are hardly needed in cells, because the cell interior contains few nuclei for crystal formation. They can, however, prevent ice crystals in the vasculature from damaging the blood-brain barrier. Cryoprotectants often only pass through the small vessels in small quantities, and the risk of ice formation there is then great.

Ice blockers almost seem to be the solution to all problems, but that would be .....

7.5.1 Too good to be true.

If ice blockers are used without the necessary concentration of other antifreeze agents or if the temperature is lowered too slowly, this can lead to ice formation at low temperatures, which is more dangerous than that at higher temperatures. Ice blockers, after all, prevent ice formation outside the cells. Because less water is then extracted from the cells by ice formation outside the cells, in such cases ice formation occurs in the cells, into which the ice blockers cannot penetrate.

The addition of ice blockers to solutions of other antifreeze agents resulted in several good successes (Capicciotti et al. 2015; Eisenberg et al. 2012; Marco-Jimenez et al. 2012).

At least in the case of VM-1 – the vitrifying solution from Cryonics Institute – a reduced "stability" of the brain vitrification, which can lead to devitrification (see below) and ice formation, is produced by addition of components such as trehalose or sucrose. These are used to reduce more harmful cryoprotectants. This could be shown on brain slices (Pichugin 2007). Thus, VM1 is more damaging than solutions with ice blockers, but vitrifies more stably. Cryonics Institute uses this solution, while Alcor uses mixtures with ice blockers such as M22 (review Best B Vitrification in cryonics https://www.benbest.com/cryonics/cryonics.html).

Another group of antifreeze agents are organic substances with small molecules (osmolytes) such as hydroxyectoin, ectoin and L-proline, which are formed by microorganisms under stress. They are compatible solutions. They have been used to reduce the amount of membrane penetrating cryoprotectants (which are often harmful due to high concentration of small molecules or specific properties) during slow freezing. Synthetic such substances are also favorable for vitrification (Fahy, Wowk 2015; Fahy et al. 2004; Fahy et al. 2013; Freimark et al. 2011; Guan et al. 2013; Leather et al. 1993; Lee, Denlinger 1991; Nickell et al. 2013; Sei et al. 2002; Sformo et al. 2010; Sun et al. 2012; Tan et al. 2012; Ting et al. 2012; -2013; Wowk, Fahy 2002; Wowk et al. 2000; Fahy, Wowk 2015; cited in Bojic et al. 2021).

7.6 Clathrates: Water locks guest molecules in cages, ice must stay out — another chance?

Clathrates could be an interesting alternative for the prevention of ice formation.

Many molecules without electrical charge (apolar molecules) can be dissolved in water without electrical attractive forces (as in hydration by bound water) being involved.

Non-electric groups in proteins include the methyl group (for example, in the amino acid alanine), the benzyl group (for example, of phenylalanine), or the isopropyl group (for example, of valine). Most of them are found on the surface of a protein. They bind by so-called hydrogen bonds which are relatively weak. Others are forced into the interior of a large protein molecule because of a certain repulsion by water (hydrophobic forces).

As the temperature decreases, this repulsion decreases and the hydrogen bonds become stronger.

Well, through all this chemistry, water molecules form a kind of cage (clathrate) around water-repellent (non-polar) "guest molecules". Thereby the water molecules themselves form a hydrate by electrical forces in which the "water-shy" molecules are enclosed by so-called van der Waals forces.

Alcohols can both occur as guest molecules or hinder clathrate formation. This depends on temperature and pressure.

Clathrates have very interesting positive effects. For example, the formation of clathrate prevents water from separating from the solutions of the tissues in the form of pure ice.

Clathrates like that of xenon could prevent damage from high salt concentration.

Clathrates can protect cells from desiccation by retaining water in the cell. This is important during freezing, where water normally leaks out of the cell into the environment. The cell is protected from damage and less ice forms outside the cell.

Sensationally, xenon clathrates are thought to protect the cells from forming ice crystals again during thawing (Alavi et al. 2010; Makiya et al. 2010).

However, clathrates do not only have advantages. The volume of a clathrate cage is larger than a hexagonal ice model with the same number of molecules, even when the guest molecules are removed. Clathrates are therefore likely to produce more damage in biological tissue than freezing (the ice crystal formation).

Clathrate-forming gases such as xenon, when released, can form bubbles that impede circulation (Pulver et al. 2013)

Clathrates are similar to ice crystals and their formation is prevented during vitrification, like that of ice crystals.

The main disadvantage of clathrates is their large volume and the mechanical damage caused by their formation (see also Best B vitrification in cryonics. https://www.benbest.com/cryonics/vitrify.html).

In any case, more research is needed for their use in antifreeze.

Literature

Alavi S et al (2010) Effect of guest-host hydrogen bonding on the structures and properties of clathrate hydrates. Chem-A Eur J 16:1017-1025

Anchordoguy J et al (1987) Modes of interaction of cryoprotectants with membrane phospholipids during freezing. Cryobiology 24:324-331

Angell CA, Sare EJ (1970) Glass-forming composition regions and glass transition temperatures for aqueous electrolyte solutions. J Chem Phys 52:1058-1068

Bai Y et al (2001) State diagram of apple slices: glass transition and freezing curves. Food Res Internat 34:89-95

Baudot A et al (2000) Glass-forming tendency in the system water-dimethyl sulfoxide. Cryobiology 40:151-158

Benson JD et al (2012) Analytical optimal controls for the state constrained addition and removal of cryoprotective agents. Bull Math Biol 74:1516-1530

Best BP (2015) Cryoprotectant toxicity: facts, issues, and questions. Rejuvenation Res 18:422-436

Bojic S et al (2021) Winter is coming: the future of cryopreservation. BMC Biol 19, 56

Brake NC, Fennema OR (1999) Glass transition values of muscle tissue. J Food Sci 64:10-15

Capicciotti CJ et al (2015) Small molecule ice recrystallization inhibitors enable freezing of human red blood cells with reduced glycerol concentrations. Sci Rep 5:9692

Courbiere B et al (2006) Cryopreservation of the ovary by vitrification as an alternative to slow-cooling protocols. Fertil Steril 86:1243-1251

Crevel RW et al (2002) Antifreeze proteins: occurrence and human exposure. Food Chem Toxicol 4:899-903

Davies PL, Hew CL (1990) Biochemistry of fish antifreeze proteins. FASEB J 4:2460-2468

De Wolf A, de Wolf C (2013) Human cryopreservation research at advanced neural biosciences. In Sames KH (ed) Applied Human Cryobiology, vol. 1, ibidem, Stuttgart, pp 45-59

Eisenberg DP et al (2012) Thermal expansion of the cryoprotectant cocktail DP6 combined with synthetic ice modulators in presence and absence of biological tissues. Cryobiology 65:117-125

Fahy GM, Wowk B (2015) Principles of Cryopreservation by Vitrification. In Wolkers WF, Oldenhof H (eds) Cryopreservation and freeze-drying protocols. Methods Mol Biol 1257 Springer Protocols Humana Press, Totowa, pp 21-82

Fahy GM et al (1990) Cryoprotectant Toxicity and Cryoprotectant Toxicity Reduction: In Search of Molecular Mechanisms. Cryobiology 27:247-268

Fahy GM et al (2004) Improved vitrification solution based on the predictability of vitrification solution toxicity. Cryobiology 42:22-35

Fahy et al (2004a) Cryopreservation of organs by vitrification: perspectives and recent advances. Cryobiology 48:157-178

Fahy GM et al (2013) Cryopreservation of precision-cut tissue slices. Xenobiotica 43:113-132

Feldschuh J et al (2005) Successful sperm storage for 28 years. Fertil Steril 84:P1017.e3-1017.e4

Fennema OR (1996) Food Chemistry, Marcel Dekker, INC. New York (3rd ed. CRC Press Inc, Boca Raton).

Freimark D (2011) Systematic parameter optimization of a Me(2)SO- and serum-free cryopreservation protocol for human mesenchymal stem cells. Cryobiology 63:67-75

Galhuber M et al (2021) Simple method of thawing cryo-stored samples preserves ultrastructural features in electron microscopy. Histochem Cell Biol 155:593-603

Gautam SK et al (2008) Effect of type of cryoprotectant on morphology and developmental competence of in vitro-matured buffalo (Bubalus bubalis) oocytes subjected to slow freezing or vitrification. Reprod Fertil Dev 20:490-496

Goff HD (1995) The use of thermal analysis in the development of a better understanding of frozen food stability. Pure Appl Chem 67:1801-1808

Guan N et al (2013) Analysis of gene expression changes to elucidate the mechanism of chilling injury in precision-cut liver slices. Toxicol In Vitro 27:890-899, doi: 10.1016/j.tiv.2012.10.009

Guerreiro BM et al (2016) Physicochemical analysis of antifreeze properties in chemical compounds and proteins for cryopreservation. B.Sc. Thesis, Lisbon

Hopkins JB et al (2012) Effect of common cryoprotectants on critical warming rates and ice formation in aqueous solutions. Cryobiology 65:169-178

Izutsu K et al (2009) Freeze-drying of proteins in glass solids formed by basic amino acids and dicarboxylic acids. Chem Pharm Bull 57:43-48

Karlsson JEM, Toner M (1996) Long-term storage of tissues by cryopreservation: critical issues. Biomaterials 17:243-256

Kashuba CM et al (2014) Rationally optimized cryopreservation of multiple mouse embryonic stem cell lines: I--Comparative fundamental cryobiology of multiple mouse embryonic stem cell lines and the implications for embryonic stem cell cryopreservation protocols. Cryobiology 68:166-175

Kawahara H (2008) Cryoprotectants and ice-binding proteins. In Margesin R et al (eds) Psychrophiles: from biodiversity to biotechnology. Springer, Berlin, pp 229-246

Kilbride P et al (2016) Spatial considerations during cryopreservation of a large volume sample. Cryobiology 73:47-54

Kheirabadi BS, Fahy G (2000) Permanent life support by kindneys perfused with a vitrifiable (7.5 molar) cryoprotectant solution. Transplantation 70:51-57

Leather SR et al (1993) The ecology of insect overwintering. Cambridge University Press, Cambridge

Lee RC (2002) Cytoprotection by stabilization of cell membranes. In Sipe JD et al (eds) Reparative medicine. Growing tissues and organs. Ann NY Acad Sci 981:271-275

Lee RE, Denlinger DL (1991) Insects at Low Temperature. Chapman and Hall, New York

Lemler J et al (2004) The arrest of biological time as a bridge to engineered negligible senescence. Ann NY Acad Sci 1019:559-563

Luechtefeld T et al (2018) Machine learning of toxicological big data enables read-across structure activity relationships (RASAR) outperforming animal test reproducibility. Toxicol Sci 165:198-212

Makiya T et al (2010) Synthesis and characterization of clathrate hydrates containing carbon dioxide and ethanol. Phys Chem Phys 12:9927-9932

Marco-Jimenez F et al (2012) Effect of "ice blockers" in solutions for vitrification of in vitro matured ovine oocytes. Cryo Letters 33:41-44

Mazur P (1963) Kinetic of water loss from cells at subzero temperatures and the likelihood of intracellular freezing. J Gen Physiol 47:347-369

Mazur P (1990) Equilibrium, quasi-equilibrium, and nonequilibrium freezing of mammalian embryos. Cell Biophys 175:53-92

Mehl PM (1993) Nucleation and crystal growth in a vitrification solution tested for organ cryopreservation by vitrification. Cryobiology 30:509-518

Naitana S (1997) Polyvinyl alcohol as a defined substitute for serum in vitrification and warming solutions to cryopreserve ovine embryos at different stages of development. Anim Reprod Sci 48:247-256

Nawroth F et al (2002) Vitrification of human spermatozoa without cryoprotectants. Cryo Letters 23:93-102

Nickell PK et al (2013) Antifreeze proteins in the primary urine of larvae of the beetle Dendroides canadensis. The J Exp Biol 216:1695-1703

Pichugin Y (2006a) Problems of long-term cold storage of patients' brains for shipping to CI. The Immotalist 38:14-20

Pulver A et al (2018) Combined approach to the development of protocol for vitrification of bulky biological objects. In Sames KH (ed) Applied Human Biostasis, vol. 2, ibidem, Stuttgart, pp 47-55

Regand A, Goff HD (2006) Ice recrystallization inhibition by ice structuring proteins from winter wheat grass. J Dairy Sci 89:49-57

Sames K (1990) Age related changes of morphological parameters in hyaline cartilage. In Robert L, Hofecker G (eds) The theoretical basis of aging research, vol. 2, Facultas, Vienna, pp 177-184

Sames K (1994) The role of proteoglycans and glycosaminoglycans in aging. In Hahn HP (ed) Interdiscip Top Gerontol, vol. 2, Karger, Basel

Schreuders PD et al (1996) Characterization of intraembryonic freezing in anopheles gambiae embryos. Cryobiology 33:487-501

Sei T et al (2002) Growth rate and morphology of ice crystals growing in a solution of trehalose and water. J Cryst Growth 240:218-229

Sformo T et al (2010) Deep supercooling, vitrification and limited survival to -100°C in the Alaskan beetle Cucujus Clavipes Puniceus (Coleoptera: Cucujidae) Larvae. J Exper Biol 21:502-509

Shaw JM et al (1997) Vitrification properties of solutions of ethylene glycol in saline containing PVP, Ficoll, or dextran. Cryobiology 35:219-229

Shinde P et al (2019) Freezing of dendritic cells with trehalose as an additive in the conventional freezing medium results in improved recovery after cryopreservation. Transfusion 59:686-696

Solanki PK et al (2017) Thermo-mechanical stress analysis of cryopreservation in cryobags and the potential benefit of nano warming. Cryobiology 76:129-139

Solocinski J et al (2017) Effect of trehalose as an additive to dimethyl sulfoxide solutions on ice formation, cellular viability, and metabolism. Cryobiology 75:134-143

Sputtek A (1996) Cryopreservation of blood cells. In Müller-Eckhard C (ed) Transfusion medicine, basics, therapy, methodology. 2nd ed. Springer, Berlin, pp 125-135

Sun WQ (1999) State and phase transition behaviors of Quercus rubra seed axes and cotyledonary tissues: relevance to the desiccation sensitivity and cryopreservation of recalcitrant seeds. Cryobiology 38:372-285

Sun H et al (2012) Compatible solutes improve cryopreservation of human endothelial cells. Cryo Letters 33:485-493

Tan X et al (2012) Successful vitrification of mouse ovaries using less-concentrated cryoprotectants with Supercool X-1000 supplementation. In Vitro Cell Dev Biol Anim 48:69-74

Tas RP et al (2021) From the freezer to the clinic. EMBO Rep 22:e52162

Ting AY et al (2012) Synthetic polymers improve vitrification outcomes of macaque ovarian tissue as assessed by histological integrity and the in vitro development of secondary follicles. Cryobiology 65:1-11

Ting AY et al (2013) Morphological and functional preservation of pre-antral follicles after vitrification of macaque ovarian tissue in a closed system. Hum Reprod 28:1267-1279

Uhlmann DR (1972) A kinetic treatment of glass formation. J Non-Cryst Solids 7:337-348

Valojerdi MR et al (2009) Vitrification versus slow freezing gives excellent survival, postwarming embryo morphology and pregnancy outcomes for human cleaved embryos. J Assist Reprod Genet 26:347-354

Weng L et al (2011) Molecular dynamics study of effects of temperature and concentration on hydrogen-bonding abilities of ethylene glycol and glycerol: implications for cryopreservation. J Phys Chem A 115:4729-4737

Wolkers WF, Oldenhof H (2021) Principles Underlying Cryopreservation and Freeze-Drying of Cells and Tissues. Methods Mol Biol 2180:3-25

Wowk B (2007) How cryoprotectants work. Cryonics (Alcor) 3rd Quart (2007) (www.Alcor.org)

Wowk B (2010) Thermodynamic aspects of vitrification. Cryobiology 60:11-22

Wowk B, Fahy GM (2002) Inhibition of bacterial ice nucleation by polyglycerol polymers. Cryobiology 44:14-23

Wowk B et al (2000) Vitrification enhancement by synthetic ice blocking agents. Cryobiology 40:228-236

Zachariassen KE, Kristiansen E (2000) Ice nucleation and antinucleation in Nature. Cryobiology 41:257-279

8 Remaining hurdles and surprising solutions

Summary

Cryopreservation of the legally deceased, but also of human organs, does not yet enable reanimation. The reasons are not only to be found in thermophysics. Great care is of course still required to avoid ice crystals. But it turns out that the greater difficulties occur during rewarming. It can put the success in question. And there are the stress cracks, the chilling injury, the cold shock. What are the remedies for this? One thing is certain: large objects also have the biggest problems.

8.1 How cells are fed, mass transfer between blood and cells

When blood flow stops, two opposite reactions can occur. On the one hand, the blood vessels can become clogged, and on the other hand, the vessel walls can become "leaky" (permeable). In the case of circulatory arrest and loss of the energy sources that reach the cells via the blood, important life processes of the cells fail. The inner wall cells of the blood vessels are also affected. They swell and vessels are obstructed as a result.

Tissue swelling, on the other hand, occurs especially when the blood vessels are damaged, for example, when the vessel walls become more permeable due to a stop in blood flow and its consequences and also allow large molecules to pass through. This is especially important in the brain. Water can leak into the tissues and cells, and can lead to brain swelling.

Normally, 67% of the fluids are in cells, 26% outside the cells and 7% in the blood. All fluids are transported via the vessels passing the vessel wall, the matrix around the cells, entering cells and flowing back again in a steady flow.

8.1.1 Role of the cell membrane as the "mouth" of the cell

The migration of particles with electric charge is more or less blocked at membranes. The cell membrane protects the interior of the cell with its metabolic chemistry from the effects of the environment. However, this also blocks the supply of substances for metabolism, the useful products of metabolism and its waste products can only pass into or out of cells to the extent that they are able to migrate through the membrane. The membrane itself is partially permeable (semipermeable).

To ensure that the cell membrane does not represent an insurmountable barrier and that a reaction of the cell with its environment remains possible, there must be passages. One could compare this with the Roman Limes, which was not a total barrier but rather served to control transport and exchange. There are channels in the cell membrane for this purpose. These allow special substances to migrate across the membrane, even without having to penetrate it.

Channels made up of proteins existing in the cell membrane allow some substances to pass through quickly (for example water, sodium ions (Na+) or potassium ions (K+).

Whether a cell in a solution shrinks due to loss of water depends on the properties of the cell membrane and, as discussed, on the osmotic pressure (tonicity) in the cell.

Cell membranes consist of a double layer of phosphorus-containing lipid molecules interspersed with built-in protein molecules. Therefore, fat-soluble substances pass more easily through the membrane (e.g., oxygen, nitrogen, carbon dioxide, and alcohols), while molecules with positive or negative charges are more likely to be blocked. However, molecules can also move through the channels that pass through the cell membrane. For example, with their help, water can quickly pass through the cell membrane. The opening of the channels can be controlled by cells. The protein molecules of the channels are not the only ones in the cell membrane. Proteins with different effects sit on the membrane or pass through it (a trick to connect inside and outside without gaps).

First, however, a substance must pass from the blood into the environment of the cell. If a substance is strongly enriched in the blood, it migrates (diffuses) — if possible — through the vascular wall into the tissue. If a substance is more enriched in the tissue than in the blood, it can, conversely, migrate through the vessel wall into the blood — in accordance with the laws of osmosis — in order to balance the concentration.

In cryonics, one now gives very highly enriched solutions of cryoprotectant because they are supposed to migrate more readily into the tissues and cells.

If migration is not possible, an osmotic pressure develops (see above). This attracts water, for example, or other substances migrate to compensate. Each molecule whether small or large contributes in the same way to the osmotic pressure. As mentioned, the osmotic pressure for a substance consisting of small molecules is therefore higher than the osmotic pressure of a substance with large molecules, if both are present in the same quantity (e.g. have the same weight). Strictly speaking, the number of particles (molecules as well as ions) is decisive for the osmotic pressure. If a molecule like NaCl can release more than one ion, the number of its ions is responsible for the osmotic pressure.

The so-called osmolarity is the number of particles in one liter of solution. In contrast, osmolality is the number of particles in one kilogram of a solution.

The walls of blood vessels also consist of cell membranes, but also of other membranes. The blood-brain barrier consists of adjacent membranes of cells and intervening membrane material. Other boundaries between cells and the bloodstream are built according to a similar principle, e.g. in alveoli and endocrine glands.

Especially for the transport of amino acids, which serve as transmitter substances of nerve excitation, the crossing of the blood-brain barrier is normally limited so that no flooding of the brain with these substances occurs (Bernacki et al. 2008). This applies, for example, to the transmitter substance norepinephrine from the adrenal gland.

Other substances that play a role in the brain are regulated in the same way. Water also crosses the barrier with greater difficulty than other hair vessel walls. This is also true for various ions (Smith, Rapoport 1986; Pardridge 2003; -2005). Thus, the blood-brain barrier is a particular control point and is of great interest to cryonics because its preservation is central to the brain.

Substances from the bloodstream thus generally migrate into the tissues and cells via the walls of the capillaries. In the brain, however, the inner wall cells of the vessels are tightly joined without interstices. This affects the transport of substances across the blood-brain barrier. This barrier is particularly sensitive to a stop in blood flow, and when flow resumes following a stop, permeability across the interstitial spaces of the cells increases (Witt et al. 2003).

Increased permeability is desired in cryoprotectant perfusion. Coincidentally, DMSO is not only antifreeze, but has also been used (along with ethanol and detergents) in artificial perfusion to open the blood-brain barrier (Pardridge 2005). Thus, this effect is also present in various vitrification solutions. Likewise, capsaicin and the contrast agent Optison have been used to break down the blood-brain barrier (Hu et al. 2005; Mychaskiw et al. 2000). Although the current approaches ensure sufficient saturation of the brain with cryoprotectant, the passage of the blood-brain barrier is difficult to control. It would have to be opened selectively without causing brain swelling.

On the other hand, the blood-brain barrier undergoes damage by stop of blood circulation. Repair mechanisms are still an object of research (Späth 2929; McMahon, Ichida 2922):

In vessels during life there is not only osmotic pressure, but also dynamic blood pressure. In the capillaries however, the blood pressure on the venous side leading back to the heart reaches zero. Blood can only flow to the heart because venous valves prevent backflow and the heart exerts suction. In the capillaries, there is still a higher pressure at their beginning on the arterial side. This causes fluid to leak into the tissue, which is collected again by lymphatic vessels and transported to the veins.

Due to the division into capillaries, the diameter of all blood vessels taken together increases. Thus, the blood pressure drops due to the resistance of the vessels.

In artificial flow (e.g., perfusion in cryonics), typically 1 or 2 L/per minute flows at a pressure of 80mm HG. If the viscosity of the solution increases (for example, due to cooling), the pressure must increase to maintain the flow rate or the fluid will flow more slowly. If the vessels (with enhanced rigidity in the cold) were to burst under too high a pressure, perfusion could fail.

With suitable concentrations of polymers such as PVP, K360, dextran 500 and dextran sulfate 500, the viscosity can be increased.

Lowering the temperature also increases the viscosity.

However, increased flow pressure in brains 24 and 48 hours of cold blood flow arrest worsened *away* the outcome (de Wolf, de Wolf 2013).

The solidification temperature of glycerol is -90°C. So, when transported in dry ice at -78°C, the glycerol was still fluent, but glycerol is viscous enough at this temperature to keep a patient in dry ice for a few days.

At 37°C, glycerol is almost 600 times more viscous than water; at 10°C, of course, it is much more viscous.

Fortunately, new cryoprotectants are much less viscous than the glycerol used in the past (see also Best B Perfusion and diffusion in cryonics protocol https://www.benbest.com/cryonics/protocol.html).

8.2 Obstacles to the storage of cells and tissues

In addition to the aforementioned obstacles in cooling, heating and freeze protection, there are others such as chilling injury and cold shock, as well as difficulties in the practical implementation of cryonics.

8.2.1 Feasibility problems with vitrification and rewarming

A review of the harms of preservation and restoration of human bodies has been provided by Ben Best (Best 2018). Perhaps the most important biological obstacle to cryonics is that resuscitation does not make sense until aging changes and disease sequelae can be repaired, a lengthy endeavor.

Disease and aging changes can only — perhaps — be eliminated in the future. The main problem with the practical approach is thus that cryonics — as already mentioned — still has to wait for legal death today, together with aging and disease damage. Practical obstacles for cryonics have been compiled by A. de Wolf (2018).

Important are: "cold shock," pH shifts, phase transformations and separations in membranes, lipid peroxidation, vesicle formation, protein changes, effects of free radicals, so-called thermoelastic stress, embrittlement, cracking, devitrification, and crystallization upon heating (see in Sputtek 1996).

We cannot — fortunately for the reader — discuss all these processes in every detail in this book. An umbrella term is: cell damage. Some of the changes have a common cause and not each requires its own countermeasures.

Surprisingly, cryonics nevertheless already works on cells and small organs.

The older method of protecting cell cultures using glycerol shows that it is mainly the concentration that makes antifreeze solutions harmful. While glycerol is present in healthy cells as an important component of sugar metabolism without being harmful at normal concentrations, the high concentrations used for antifreeze are harmful. By the way: almost all substances in our food are also toxic in very high concentrations. Cooling reduces harmful effects (Fahy et al. 1990; -2004; Wowk et al. 2000). Therefore, the damage can be reduced by shortening the exposure to the sample, by dilution or by lowering the temperature, or mixtures can be used instead of pure solution. Ideally, the temperature should be reduced to as low as +10 to -4°C before the highest necessary concentrations of cryoprotectant are applied.

When rewarming culture cells, for example, one ensures rapid attainment of the liquid state and rapid dilution, and this has worked routinely on cell cultures for a long time.

The harmful effect of cryoprotectants becomes more dangerous again at higher temperatures. The difficulties of rewarming are discussed below.

8.3. Problems with object size, chilling injury and cold shock

One must not see an organ as a uniform block. We cool via the circulation as long as the flow can be maintained. Since the capillaries come within nanometers of the most important cells, the layers between vessels and cells are very thin, and we turn off the pump if possible not before cryoprotectant has reached everywhere. The pump is switched off, however, when the cryoprotective solution becomes too viscous due to cooling. Now cooling can no longer take place via the circuit, but only from the outside.

During freezing, however, as with vitrification, the entire tissue then becomes a solid block. It is only cooled or heated from the surface. Inside, the temperature change lags behind. This leads to stresses and cracks. Cracks occur if cooling continues once the glass transition temperature has been reached (Adam et al. 1990, and see below). Even samples as small as a few milliliters show cracks when cooled to the temperature of liquid nitrogen.

Breaking occurs mainly during rapid cooling as well as heating (see below).

Unfortunately, vitrified cryoprotectant solutions are quite sensitive to thermodynamic stress. Today, one can only counteract this to a limited extent by stopping the cooling just before the glass transition temperature is reached for a longer period of time so that the temperatures can equalize between the exterior and interior. In addition, cooling from the glass transition temperature down to the temperature of liquid nitrogen is carried out very slowly. In addition, ice blockers show some effect against cracking. They reduce the extent of freeze fractures, but do not prevent them completely.

So, overall, the major disadvantages of vitrification are the toxic effects, breaking or bursting, and the possibility of ice formation within the cell due to rapid cooling. According to the research organization 21Century Medicine, the upper limit of damage in vitrification is, 2% ice crystal formation.

Whether it is possible to delay cooling so that cracks do not occur is almost impossible to test in humans for practical reasons. It could simply take too long in practice with such a large body.

Today, when human bodies are frozen, the cracks are not repaired. This, like the elimination of aging changes and disease consequences, is left to the future. Repairing a crack is likely to be a simpler problem than, for example, stopping and eliminating aging changes. However, in animal experiments or in the cryopreservation of transplant organs in which resuscitation is attempted, this problem must currently be addressed.

A nearly ideal possibility would be storage at temperatures close to the glass transition temperature, but the temperature must not be higher, because above -138°C large ice districts increase at the expense of the smaller (more harmless) ones. This makes storage at -80°C, for example, harmful for longer than half a year. One must therefore store at temperatures just below -138- 140°C (Petrenko et al. 1999). Against this, perhaps, speaks the survival of tiny creatures for about 20,000 years at the relatively high polar temperatures, as we discuss below.

The cryonics organization Alcor made a test with storage at temperatures not far below the glass transition temperature.

Cracks only occur below the glass transition temperature. The lower the temperature below the glass transition temperature, the more intense the freeze fractures. Unfortunately, the available technology for storage at temperatures above -196°C is not quite as simple as storage in liquid nitrogen. Further, storage at correspondingly high temperatures is expensive (Best 2013; -2013b; Fahy et al. 1990; -1990a; Wowk 2011).

At Cryonics Institute (CI) in Michigan, they are now actually going down to nitrogen temperature with VM1 vitrification solution, which makes continuous cooling easier while increasing the problem of cracking. That is why they are cooling quickly to -120°C with VM1 cryoprotective solution and then slowly over about 5 days down to liquid nitrogen temperature to minimize fracture damage.

The limit for an acceptable deviation in the cooling rate between the outside and inside of a specimen is about 1 mm layer thickness. So far only for pieces with edge lengths below one millimeter the differences between inside and outside are too small to produce cracks.

In 1990 G. Fahy et al. published results of experiments on freeze fractures with propylene glycol as cryoprotectant. Propylene glycol has a glass transition temperature of -108°C, but dissolved in the carrier solution RPS-2 it is -102°C. Uniformity and rate of cooling are crucial determinants for the development of fractures. In one experiment, freeze fractures for smaller samples started at lower temperatures: -143°C for a 46ml sample, -116° for 482ml, and -111° for 1412ml the last volume comparable to that of a human brain.

At the lower temperatures, the cracks were thinner and more numerous than at higher temperatures. The lower the temperature was below the glass transition temperature, the more severe and finely distributed were the freeze fractures (Rabin, Plitz 2005). Large cracks occurred due to faster cooling and at higher temperatures.

Alcor used a drastic method to detect cracking during cooling of patients. They placed a listening device ("crackphone") under the top of the skull to hear the cracking.

Biological objects fortunately resist breaking better than pure solution (with which the preliminary tests are often made).

One question, however, remains open, namely whether there is a stronger mobility of molecules when stored at -140°C instead of in liquid nitrogen (Best 2013b). So, storage at -196°C might be more favorable in the end.

Near the glass transition temperature, cooling can be slower to avoid the surface freeze fractures because the viscosity is already too high for ice formation.

Objects the size of a human head cannot be kept more than 20°C below the glass transition temperature without freeze fracture.

Storage at temperatures just below the glass transition temperature reduces freeze fractures and formation of crystallization nuclei, but does not prevent them completely.

Increasing the thermal conductivity could possibly prevent the fractures.

Unfortunately, even with -130 °C storage, ruptures can occur where perfusion does not reach all tissue areas. In addition, it is not clear how long one can really store at these temperatures. For example, radiation can also act more easily at such temperatures.

Liquid nitrogen is currently simply the safest and least expensive medium for storing frozen biological objects (see also Best B Quantifying Ischemic Damage for Cryonics Rescue https://www.benbest.com/cryonics/IR_Damage.html).

8.3.1 Long distances, uneven structures

In large objects, properties in the structure of the tissues are added to the large distances for exchange processes. Mutual dependencies of structures exist in many respects e. g. cross-linking of cells with other cells and of cells with matrix substances as well as vice versa. In the same organ there are different types of cells and tissues (working tissue, nerves, vessels). The matrix between cells also can be harmed, for example, during slow freezing (Schenke-Layland et al 2007). In the case of artificial tissues, properties of exogenous materials are added. If the heat distribution within the tissue differs, fractures occur more easily. Water can remain in blood vessels, freeze there and lead to rupture. All this can lead to inequalities in temperature changes.

Delays in flow or temperature change in thick layers, of course, also increase the risk that harmful deterioration will accumulate (Baxter, Lathe 1971). It is also difficult to distinguish the consequences of stresses due to vitrification and rewarming and the damaging effects of cryoprotectants. In addition, there are methodological variations for the specific cell types (McLellan, Day 1995).

What would be required above all is a higher rate cooling and heating in the interior of the specimen.

The migration and penetration rates of molecules are additionally lowered in tissues at low temperature. Inside the organs self-digestion of the tissue can start early when the blood circulation is interrupted. The tissue complex, unlike cell cultures, also opens up possibilities for damage to the cell connections, etc.

Inside a large organ, the temperature change may be so delayed that during cooling the tissue not yet saturated with antifreeze suffers damage.

With a volume that is large compared to the surface area, the vitrification solutions consequently should require a higher starting concentration to arrive inside at an effective concentration. However, this is not a major problem in cooling as long as perfusion is ongoing. It is interesting to find solutions that allow much slower cooling and heating without creating ice crystals. In this way, one can give time to the perfusion and work with lower concentrations. This was shown for the DP6 solution and various additives further enhanced this property (Wowk et al. 2018).

Capillaries in tissue specimens apparently do not improve the migration of substances without flow, e.g. in isolated tissue specimens which are not connected to the blood circulation. These are therefore of limited use to study the application of cryoprotective solutions. In a 1 cubic centimeter sample, a delay of 1 hour was observed for DMSO uptake in the interior of the sample (porcine cardiac muscle tissue) compared to the surface. This makes damage to superficial cells almost impossible to avoid if cooling is delayed until the necessary concentration of cryoprotectant is reached inside.

However, if many capillaries are present in the tissue, the process is accelerated as long as the cryoprotective solution flows in them. As mentioned above, capillaries come close to the cells in tissues with active metabolism, usually at distances far below the millimeter range. Therefore, heat exchange and the supply of cryoprotective substances is favorable for the cells as long as the solution is flowing.

In 2020, Bleisinger et al. reported the observation of perfusion using an iodine-containing contrast agent on rat hearts. They observed that DMSO at room temperature diffused the cells and extracellular spaces up to a 95% exchange in 35 seconds. The subsequent washout required 49 seconds. The authors advise shortening the perfusion with toxic cryoprotectants at temperatures above the freezing point. Thus, tissue layer thickness does not appear to be an obstacle to the uptake of the cryoprotectant during perfusion.

Since the cryoprotective solutions are already offered above the freezing point of water, the distances they have to overcome are small, namely only from the capillaries to the cells. Only when the flow slows down and stops at low temperatures do delays increase.

Then, ideally, the cryoprotectants should already have arrived in the cells at temperatures at which their harmful effect has already greatly diminished.

The delays inside large samples are therefore a problem particularly during thawing and the consequent dilution of harmful cryoprotectant concentrations.

Delays, by the way, are not just inconvenient. They mitigate too rapid concentration or dilution. This also reduces to rapid a change in osmotic pressure and the damage that can be associated with it.

Fahy and coworkers have summarized (1990; 1990a) the problems associated with large-body cooling.

Facing all difficulties, research on large objects is necessary. E.g. isolated cells react differently than cells in living tissue, especially after a longer time in culture. This concerns e. g. permeability of membranes, water removal and crystallization (Karlsson, Toner 1996). It stands to reason that cell cultures are unsuitable as a sole model for cryobiology. This underscores our view that research should also be done on organs, on animals, and even on deceased humans.

8.4 Chilling injury and cold shock

These phenomena reduce the viability of cells at low temperatures, even above freezing, but not so high that the cells of warm-blooded animals function normally.

Chilling injury in animal cells likely results from a phase transition in cell membranes (Hays et al. 2001).

Cold shock is manifested by reduced cell viability. It occurs either through a rapid or a deep drop in temperature (Al-Fageeh, Smales 2006). The structure of the nucleic acids DNA and RNA and protein molecules that react with them play a role in this process. Cold shock has been most carefully studied in bacteria. It also affects almost directly the lipids of the membranes (Weber, Marahiel 2002). The passage of substances through the cell membrane is thereby reduced.

There is an overlap in the effects of cold shock and chilling injury on cell organelles, especially at the membranes. Unfortunately, the terms also are somewhat confused.

Fatty substances in cell membranes are the molecules that repel water. They are assumed to transition from the liquid phase to a viscous semi-solid gel phase during (freeze-fire) cooling between 0- and -20°C. In this temperature range, chilling injury is most pronounced (Murata et al. 1992). Thereby permeability of membranes increases due to irregularities in the packing of molecules between liquid phase and this semi-solid gel phase.

> Background information
> More important is probably a reaction between protein molecules in the membranes and fatty substances located there that contain phosphorus compounds (phospholipids), which have something to do with the transition between the phases.

In his vitrification experiment with ethylene glycol on embryos of the fly Drosophila (fruit fly), which already possess 50,000 cells, cryobiologist Mazur found that the embryos were extremely sensitive to chilling injury Such embryos already possess tissues and organs such as muscles and nerves. For these embryos, underrunning of chilling injury by cooling at 20 000 °C per minute was essential for successful vitrification (Mazur et al. 1992; -1992a). Free radical damage during chilling injury has also been demonstrated in houseflies (Rojas, Leopold 1996).

Fish embryos cannot be cryopreserved even by such rapid cooling (Liu et al. 2001).

Platelets are also particularly vulnerable to chilling injury and thus adult mammalian tissue is also affected (Gousset et al. 2004).

Another damage caused by chilling injury and cold shock is the alteration (denaturation) of proteins. Some of the proteins affected are, of all things, antioxidant enzymes that neutralize dangerous oxygen compounds and free radicals. Types of the enzyme superoxide dismutase (SOD) are particularly concerned. But the antioxidant enzyme catalase also shows a decrease in its activity when the temperature is lowered. These enzymes play an essential role in dealing with free radicals.

When cortical sections from rat kidney are cooled in vitrification solution, chilling injury increases linearly from 0°C to -85°C. This cannot be explained by a phase transition of the membranes. Here, cell viability is the measure (Fahy et al. 2004). The K/Na ratio is decreased.

The ratio of K/Na ions in the cell indicates if the K/Na pumps are working and the cell membranes are intact. The pumps function only if enough high-energy phosphate (e.g. ATP) is produced.

For example, when the K/Na ratio is 85% (of normal) the cells are said to be 85% vital. The measurement is cheap and simple (see also Best B viability, cryoprotectant toxicity and chilling injury in cryonics. https://www.benbest.com/cryonics/viable.html).

8.4.1 Living beings protect themselves

Rapid cooling or heating reduces membrane damage. Chilling injury increases with time and can be reduced by rapid cooling in the critical temperature range (Hays et al. 2001; Mazur et al. 1992a). Vitrification can, so to speak, underrun the chilling injury (Fahy, Wowk 2021).

A higher temperature for the phase transition is found in those membranes that contain a higher proportion of saturated fatty acids.

Dehydration or the addition of a sugar also change the phase transition temperature (Koster et al. 2000).

The sugar trehalose (see above) can protect membranes against chilling injury by hydrogen bonding with their phospholipids and proteins, whereas sucrose, for example, does not (Benaroudj et al. 2001; Crowe et al. 2003).

Fortunately, there are other measures against the damage caused by chilling injury. For example, the exclusion of air can reduce chilling injury

Solutions with enhanced osmotic pressure (hypertonic solutions 1.2-1.5-fold isotonic) completely extinguished chilling injury between 0°C and -22°C. Below this temperature to -135°C, chilling injury was measured by the K/Na ratio, meaning cell survival. Vitrification solutions with excess osmotic pressure have an increased concentration of non-membrane penetrating (large) components. While the reasons for the protective effect of the increased osmotic pressure are poorly understood, it is thought that it causes cell membranes to contract less in response to heat (Fahy, Wowk 2015; Fahy et al. 2004).

Some creatures have the amazing ability to increase the saturation of fatty acids as needed when the temperature drops, and thereby to the phase transition temperature.

Plants can be insensitive to chilling injury. Their trick is an increased amount of the enzyme catalase. This enzyme — as mentioned — acts against free radicals.

Chilling injury threatens the viability of the cells more than their structure and may therefore be less of a concern in cryonics.

Evidence of a positive course is very valuable, because it says that this damage can be survived.

8.5 Daring step to success: warming up

Warming and resuscitation are already possible with small organs. In the case of the human body, apart from a longer period without oxygen supply, its size at the moment is probably the most important obstacle, especially for warming. Some causes for this are known and shall be discussed here.

8.5.1 Recrystallization, the comeback of ice crystals

A more severe problem than crystallization on cooling is the so-called recrystallization on rewarming, when the glass dissolves (devitrification) and the temperature is still below 0°C in the range of ice crystal growth.

In general, the high heat capacity of water causes delayed heat conduction, which makes warming difficult.

You can see a conglomeration of ice particles that had formed despite rapid cooling. Larger ice districts grow at the expense of smaller ones.

The crystal formation that we anxiously avoid during cooling can therefore still catch up with us during rewarming.

One reason is that the cryoprotectant concentration, which is at least necessary to achieve vitrification, is on the other hand too low for the prevention of devitrification. However, this also implies that cryoprotectants can favorably influence devitrification and the critical rate of heating (Armitage 1991; - 2002; Karlsson 2001).

Heating rates that are appropriate to the type of cells and tissues can potentially help mitigate the problem of devitrification and recrystallization, and vitrification solutions must be suitable for this (see in Fahy, Wowk 2015).

Vitrified organs, unfortunately, can also easily develop cracks during heating if it is done too quickly from the outside (Scudellari 2017).

8.5.2 How warming leads to ice

When cooling with cryoprotectant, the temperature of the strongest crystal growth is not important, because at these temperatures there are still hardly any crystallization nuclei. The main problem is the temperature of the strongest formation of crystallization nuclei. It is lower than the temperature at which the ice crystals grow and multiply (Asahina et al. 1970; Fahy, Wowk 2021). For vitrification solution M22, the temperature of greatest formation of crystallization nuclei is -110 – -120°C. It virtually ceases above -90- and below -140°C.

In contrast, the strongest growth of the crystals occurs at temperatures of -50 – -80°C and is practically terminated at -95°C (Wowk, Fahy 2007).

This is also due to the fact that crystal formation (nucleation) requires low temperatures, where it is hindered by increased viscosity, while crystal growth becomes possible as the viscosity decreases with heating.

Currently, it is considered safe to send cryonics patients from Europe to the U.S. after cooling to -78.5°C (i.e. in dry ice). There, they are then cooled to a lower temperature. Shipping in liquid nitrogen, on the other hand, would bring the transport containers to the weight limit of the airlines.

When heating thick vitrified tissue layers, the heating arrives inside with a delay and this can lead to crystallization. Some crystals may be present in the cells, which trigger crystal growth if heating is too slow. Such crystals in the cells are formed during rapid freezing which allows no sufficient dehydration of the cells.

8.5.3 Does warming challenge the total advantages of deep cooling?

The fact that crystal growth in vitrified samples can be so much more violent when heated than when cooled is called the devitrification problem.

If a sample is thawed, tiny ice particles can be formed at temperatures below 0°C and the small ice crystals, which have already formed during cooling, are ideal nuclei for crystallization. With further heating, crystal growth can then take place. During cooling, the reverse is true. Only shortly before the glass transition temperature do crystallization nuclei become available during cooling, but this comes too late for crystal growth.

Somewhat misleadingly, devitrification often does not mean the dissolution of glass when heated, but the formation of crystals during heating.

During cooling, it is ideal to be fast enough that glass formation occurs before the nuclei of crystallization take effect. Even if crystallization nuclei (small ice crystals) form, rapid cooling and the very high viscosity near the glass transition temperature prevent rapid growth of the crystals. Large ice crystals are—as mentioned—more harmful than small ones. Very small ones may even be harmless while cooling down (see also Best B vitrification in cryonics. https://www.benbest.com/cryonics/vitrify.html).

All in all, we can cool samples quite adequately today, but not yet thaw them just as well.

Accordingly, Mazur and Seki demonstrated in 2011 on a small object—the egg cell—that the success of vitrification depends less on the cooling method than on the method of heating.

8.5.4 Cryonics strategies to mitigate "devitrification".

When thawing with start at low temperatures, the formation of ice crystals is the biggest obstacle. For example, after vitrification, the heating rate would have to be 300°C/min (from -100 to 0°C in less than 20 sec) to avoid them. Very small embryos can be heated rapidly. High heating rates, if following vitrification of very small embryos with a lower concentration of a vitrification solution and slower freezing, support survival (Seki et al. 2014).

Water-repellent additives such as n-propanol or methyl-1,2-propanediol allow slower heating without increasing damage.

Recrystallization can damage slowly frozen cells from which water has been extracted during heating (probably by hasty reabsorption of water that crystallizes in the cell). Therefore, slow heating is recommended for slowly cooled samples where the water has been extracted from the cell (Karlsson 2001; Karlsson, Toner 1996) and fast heating for rapidly cooled samples.

However, the optimal fast heating rates have so far only been achieved for very small objects.

It is obvious that the critical velocities for the prevention of crystal formation are different for cooling and heating. For cooling, as mentioned, one must expect few nuclei of crystallization, for heating, many. Remarkably, many small crystals (100-3000nm) are better tolerated than a few large ones. It has been shown that a stop during heating at the glass transition temperature for some cell types can affect the influence of physical forces during dewing (Asahina et al. 1970; Fahy, Wowk 2015; Solanki et al. 2017; Takahashi et al. 1988).

In particular, the new method of electromagnetic heating of nanoparticles could lead to a breakthrough in cryonics. Manuchehrabadi et al. (2017; 2018) showed that inductive heating of magnetic nanoparticles can be used to accelerate the heating (see also Etheridge et al. 2013, Risco et al. 2018; Solanki et al. 2017). The nanoparticles must be allowed to migrate into the tissue beforehand. In principle, they can heat objects almost regardless of size. Such manipulations are not easy especially controlling the electric field.

Washing out of the nanoparticles however, is only a problem in tissue samples cut out of an organ. Using arterio-venous perfusion of an organ, leaves the particles in the capillaries and they are easily washed out by repeated perfusion (Chiu-Lam 2021; Han et al 2023).

Again, we are to point out the narrow distance between capillaries and most of the active cells in our tissues.

Focused ultrasound heating was successfully used to thaw and revive the nematode C. elegans, which had been cooled to -80°C in culture. This is excellent evidence that the method works. The focal point can be enlarged so that larger objects can be heated (Olmo et al. 2021; Risco 2021). However, these worms are very robust and can be revived from deep freezing by other means also.

Electromagnetic heating and induction have further been proposed for thawing (Evans et al. 1992; Luo et al. 2006; Robinson et al. 2002; Ruggera, Fahy 1990; Wowk, Corral 2013; Wusteman et al. 2002a; -2004). The different methods are clearly presented in Taylor et al. 2019 (see also Bojic et al. 2021).

Since the cryoprotectant solutions remain liquid below 0°C, it should in principle be possible to dilute and remove them before the organism as a whole is heated too high. Since even viscous solutions move when there is a pressure difference, perhaps one could try to allow a pressure that is not too high to act for a very long time to wash out the antifreeze at temperatures below 0°C. The most difficult thing to do here is to drive the viscous fluid across the network of capillaries, which itself forms such a high resistance that in a living state the blood pressure here drops to zero.

During thawing by microwaves (Burdette et al. 1980), disorder is created by reflux of waves at obstacles and hot and cold spots are formed. Microwave oven frequencies (2450 megahertz) cannot be used for rapid heating because of this uneven temperature distribution, as the tissues are already overheating in places, but are still susceptible to ice formation in other places. Electromagnetic waves of 300-1000 megahertz can heat more uniformly than microwaves.

Propylene glycol can be heated in a 434 megahertz di-electric field with greater uniformity than, for example, 2,3-butanediol (Robinson et al. 2002).

In Drosophila fly embryos of 50 000 cells, heating of 100 000°C/minute was required to avoid devitrification. There were 12% of embryos surviving this procedure. Even when heated at the high rate of 2000°/minute, embryos did not survive (Mazur et al. 1992).

In 2004, 21st Century Medicine developed a solution containing cryoprotectants and ice blockers that avoids devitrification when heated by 0.4°C per minute in a 10 ml sample (Fahy et al. 2004). A vitrification solution that is not concentrated high enough is metastable i.e., it will devitrify if it is not cooled or warmed quickly enough.

8.6 Premature use or experiment? Cryonics for prolonging the lives of people living today

Cryonics is often considered to be, in a narrow sense, only the cryopreservation of the patient's own body, which is done at the patient's request. It can be performed only with the immature means that we currently already have. This is cryonics after organ failure (death) and thus usually after prolonged oxygen deprivation. By today's medicine, total organ failure after a few minutes' duration can be reversed only in exceptional cases (e.g. drowning in ice water). Unfortunately, organ failure must be waited for to start human cryopreservation as long as we cannot resuscitate human bodies cryopreserved at temperatures below -130°C.

The premature application of cryonics to humans — i.e. without proving animal experiments or clinical studies of reanimation — is understandably — controversial. However, cooling can in principle temporarily stop the final cell death of the vast majority of all cells that are still alive. This alone is reason to hope for a chance of revival, however small it may be (see below, especially since we can buy new damage when cryonics is performed).

What remains is to carry out what is feasible today with the best scientifically proven methods. In the process, methods can be improved and mistakes can be learned from. In a long-term project with ongoing research and with constantly new ideas, cryonics for humans is increasingly being realized.

Methods for application to human organs and bodies are currently in the planning stage with initial steps into experimentation. Any problem identified may lead to new experiments. However, the small number of cryonics supporters worldwide results in scarce resources, so that urgent investigations fall by the wayside.

After earlier doubts, it is now accepted that small frozen organs can be resuscitated. They are composed not only of different cells, but even of different whole tissues.

The difficulties we have discussed above have to be accepted when cryonizing a deceased person already today. However, the possible solutions discussed show that even seemingly overwhelming problems are not set in stone

Literature

Adam M et al (1990) The effect of liquid nitrogen submersion on cryopreserved human heart valves. Cryobiology 27:605-614

Al-Fageeh MB, Smales CM (2006) Control and regulation of the cellular responses to cold shock: the responses in yeast and mammalian systems. Biochem J 397:247-259

Armitage WJ (1991) Preservation of viable tissues for transplantation. In Fuller BJ, Grout BWW (eds) Clinical Applications of Cryobiology. CRC Press, Boca Raton, pp 170-189

Armitage WJ (2002) Recovery of endothelial function after vitrification of cornea at -110 degrees C. Invest Ophthalmol Vis Sci 43:2160-2164

Asahina E et al (1970) A stable state of frozen protoplasm with invisible intracellular ice crystals obtained by rapid cooling. Exp Cell Res 59:349-358

Baxter SJ, Lathe GH (1971) Biochemical effects of kidney exposure to high concentrations of dimethyl sulphoxide. Biochem Pharmacol 20:1079-1091

Benaroudj N et al (2001) Trehalose accumulation during cellular stress protects cells and cellular proteins from damage by oxygen radicals. J Biol Chem 276:24261-24267

Bernacki J et al (2008) Physiology and pharmacological role of the blood-brain barrier. Pharmacol Rep 60: 600-622

Best B (2013) Cryonics: Introduction and technical challenges. In Sames KH (ed) Applied human cryobiology, vol. 1, ibidem, Stuttgart, pp 61-77

Best BP (2013b) Effects of temperature on preservation and restoration of cryonics patients. Cryonics Magazine (Institute Evidence-based Cryonics)

Best BP (2018) Forms of cryopreservation damage and strategies for prevention and mitigation. In Sames KH (ed) Applied Human Cryobiology, vol. 2, ibidem, Stuttgart, pp 75-81

Bleisinger N et al (2020) Me2SO perfusion time for whole-organ cryopreservation can be shortened: results of micro-computed tomography monitoring during Me2SO perfusion of rat hearts PLOS ONE 15:e0238519.

Bojic S et al (2021) Winter is coming: the future of cryopreservation. BMC Biol 19, 56

Burdette EC et al (1980) Microwave thawing of frozen kidneys: a theoretically based experimentally-effective design. Cryobiology 17:393-402

Chiu-Lam A et al (2021) Perfusion, cryopreservation, and nanowarming of whole hearts using colloidally stable magnetic cryopreservation agent solutions. Science Advances 7(2):eabe3005

Crowe JH et al (2003) Stabilization of membranes in human platelets freeze-dried with trehalose. Chem Phys Lipids 122:41-52

De Wolf A (2018) Identification, validation, and implementation of new cryonics technologies (an essay). In Sames KH (ed) Applied Human Cryobiology, vol. 2, ibidem, Stuttgart, pp 83-94

De Wolf A, de Wolf C (2013) Human cryopreservation research at advanced neural biosciences. In Sames KH (ed) Applied Human Cryobiology, vol. 1, ibidem, Stuttgart, pp 45-59

Etheridge ML et al (2013) 003 Radiofrequency heating of magnetic nanoparticle cryoprotectant solutions for improved cryopreservation protocols. Cryobiology 67:398-399

Evans S et al (1992) Design of a UHF applicator for rewarming of cryopreserved biomaterials. IEEE Transactions on Biomedical Engineering 39:217-225

Fahy GM, Wowk B (2015) Principles of cryopreservation by vitrification. In Wolkers WF, Oldenhof H (eds) Cryopreservation and freeze-drying protocols. Methods Mol Biol 1257 Springer Protocols Humana Press, Totowa, pp 21-82

Fahy GM, Wowk B (2021) Principles of ice-free cryopreservation by vitrification. In Wolkers WF, Oldenhof H (eds) Cryopreservation and freeze-drying protocols. Methods Mol. Biol 2180, 4th ed. Springer Protocols Humana Press, Totowa, pp 27-97

Fahy GM et al (1990) Cryoprotectant Toxicity and Cryoprotectant Toxicity Reduction: In Search of Molecular Mechanisms. Cryobiology 27:247-268

Fahy GM et al (1990a) Physical problems with the vitrification of large biological systems. Cryobiology 27:492-510

Fahy GM et al (2004) Improved vitrification solution based on the predictability of vitrification solution toxicity. Cryobiology 42:22-35

Fahy G (2013) Consequences and control of ice formation in the renal inner medulla. Cryobiology 67:409-410

Fahy G (2015) Conference abstract 16. controlling cryoprotectant toxicity and chilling injury. Cryobiology 71:169

Fuller BJ et al (eds) (2004) Life in the Frozen State CRC Press, doi: 10.1201/9780203647073

Gousset K et al (2004) Important role of raft aggregation in the signaling events of cold-induced platelet activation. Biochim Biophys Acta 1660:7-15

Han Z et al (2023) Vitrification and nanowarming enable long-term organ cryopreservation and life-sustaining kidney transplantation in a rat model. Nat Commun 14: 3407 (art. Nr.)

Hays LM et al (2001) Factors affecting leakage of trapped solutes from phospholipid vesicles during thermotropic phase transitions. Cryobiology 42:88-102

Hu DE (2005) TRPV1 activation results in disruption of the blood-brain barrier in the rat. British J Pharmacol 146:576-584

Karlsson JO (2001) A theoretical model of intracellular vitrification. Cryobiology 42:154-169

Karlsson JO, Toner M (1996) Long-term storage of tissues by cryopreservation: critical issues. Biomaterials 17:243-256

Koster KL et al (2000) Effects of vitrified and nonvitrified sugars on phosphatidylcholine fluid-to-gel phase transitions. Biophys J 78:1932-1946

Liu XH et al (2001) Effect of cooling rate and partial removal of yolk on the chilling injury in zebrafish (Danio rerio) embryos. Theriogenology 55:1719-1731

Luo D et al (2006) Development of a single mode electromagnetic resonant cavity for rewarming of cryopreserved biomaterials. Cryobiology 53:288-293

Manuchehrabadi N et al (2017) Improved tissue cryopreservation using inductive heating of magnetic nanoparticles. Sci Transl Med 9(379):eaah4586

Manuchehrabadi N et al (2018) Ultrarapid inductive rewarming of vitrified biomaterials with thin metal forms. Ann Biomed Eng 46:1857-1869

Mazur P, Seki S (2011) Survival of mouse oocytes being cooled in a vitrification solution to -196° at 95° to 70,000°C/min and warmed at 610° to 118,000°C/min: A new paradigm for cryopreservation by vitrification. Cryobiology 62:1-7

Mazur P et al (1992) Cryobiological preservation of Drosophila embryos. Science New Series 258:1932-1935

Mazur P et al (1992a) Characteristics and kinetics of subzero chilling injury in Drosophila embryos. Cryobiology 29:39-68

McLellan MR, Day JG (1995) Cryopreservation and freeze-drying protocols. Introduction. Methods Mol Biol 38:1-5

McMahon AP, Ichida JK (2022) Repairing the blood-brain barrier, engineered Wnt ligands specifically target blood-brain barrier function. Science 375: 715-771

Murata et al (1992) Genetically engineered alteration in the chilling sensitivity of plants. Nature 356:710-713

Mychaskiw G et al (2000) Optison (FS069) disrupts the blood-brain barrier in rats. Anesth Anal 91:798-803

Olmo A et al (2021) The Use of High-Intensity Focused Ultrasound for the Rewarming of Cryopreserved Biological Material. in IEEE Transactions on Ultrasonics, Ferroelectrics, and Frequency Control 68:599-607

Pardridge WM (2003) Blood-brain barrier drug targeting: the future of brain drug development. Mol Interv 3:90-105

Pardridge WM (2005) The blood brain barrier: bottleneck in brain drug development. NeuroRx 2:3-14

Petrenko VF, Whitworth RW (1999) Physics of ice. Oxford University Press (OUP)

Rabin Y, Plitz J (2005) Thermal expansion of blood vessels and muscle specimens permeated with DMSO, DP6, and VS55 at cryogenic temperatures. Ann Biomed Eng 33:1213-1228

Risco R (2021) Experimental Evidence of Return to Life with High intensity Focused Ultrasound. Biostasis the annual biostasis conference, Zurich

Risco R et al (2018) New advances in organ cryopreservation. Electromagnetic rewarming and selective targeting of ice nuclei. In Sames KH (ed) Applied human cryobiology, vol. 2, ibidem, Stuttgart, pp 65-74

Robinson MP et al (2002) Electromagnetic re-warming of cryopreserved tissues: effect of choice of cryoprotectant and sample shape on uniformity of heating. Phys Med Biol 47:2311-2325

Rojas RR, Leopold RA (1996) Chilling Injury in the Housefly: Evidence for the Role of Oxidative Stress between Pupariation and Emergence. Cryobiology 33:447-458

Ruggera PS, Fahy GM (1990) Rapid and uniform electromagnetic heating of aqueous cryoprotectant solutions from cryogenic temperatures. Cryobiology 27:465-478

Schenke-Layland et al (2007) Optimized preservation of extracellular matrix in cardiac tissues: implications for long term graft. Ann Thorac Surg 83:1641-1650

Scudellari M (2017) Core concept: cryopreservation aims to engineer novel ways to freeze, store, and thaw organs. PNAS 114:13060-13062

Seki S et al (2014) Extreme rapid warming yields high functional survivals of vitrified 8-cell mouse embryos even when suspended in half-strength vitrification solution and cooled at moderate rates to -196°. Cryobiology 68:71-78

Smith QR, Rapoport SI (1986) Cerebrovascular permeability coefficients to sodium, potassium, and chloride. J Neurochem 46:1732-1742

Solanki PK et al (2017) Thermo-mechanical stress analysis of cryopreservation in cryobags and the potential benefit of nanowarming. Cryobiology 76:129-139

Späth K (2020) Die Rolle von Osteopontin für die Reparatur der Blut-Hirn-Schranke nach experimentellem Hirninfarkt. Dissertation, Düsseldorf

Sputtek A (1996) Cryopreservation of blood cells. In Müller-Eckhard C (ed) Transfusion medicine, principles, therapy, methodology, 2nd ed. Springer, Berlin, pp 125-135

Takahashi et al (1988) Mechanism of cryoprotection by extracellular polymeric solutes. Biophys J 54:509-518

Taylor MJ et al (2019) New approaches to cryopreservation of cells, tissues, and organs. Transfus Med Hemother 46:197-215

Weber MHW, Marahiel MA (2002) Coping with the cold shock response in the Gram-positive soil bacterium Bacillus subtilis. Philos Trans R Soc Lond B Biol Sci 375:895-907

Witt KA et al (2003) Effects of hypoxia-reoxygenation on rat blood-brain barrier permeability and tight junctional protein expression. Amer J Physiol 285:H2830-H283

Wowk B (2011) Systems for intermediate temperature storage for fracture reduction and avoidance. Cryonics (Alcor) 3rd Quart 2011

Wowk B, Corral A (2013) 023 Adaptation of a commercial diathermy machine for radiofrequency warming of vitrified organs. Cryobiology 67:404

Wowk B, Fahy GM (2007) Ice nucleation and growth in concentrated vitrification solutions. Cryobiology 55:330 (Abstract 21)

Wowk B et al (2000) Vitrification enhancement by synthetic ice blocking agents. Cryobiology 40:228-236

Wowk B et al (2018) Vitrification tendency and stability of DP6-based vitrification solutions for complex tissue cryopreservation. Cryobiology 82:70-77

Wusteman MC et al (2002a) Electromagnetic re-warming of cryopreserved tissues: effect of choice of cryoprotectant and sample shape on uniformity of heating. Phys Med Biol 47:2311-2325

Wusteman M et al (2004) Vitrification of large tissues with dielectric warming: biological problems and some approaches to their solution. Cryobiology 48:179-189

9 A widely unknown success story: cryopreservation of cells embryos, tissues and small organs – many people have been "frozen" before

Summary

The fact that there are many people living among us who have already been frozen is often not realized and not well known. Frozen storage of sperm, eggs and embryos of small size is highly developed and used.

Numerous cell types, tissues and organs could be cooled below 0°C, stored there and revived. An overview is given here. The temperatures of storage range from supercooling, which is possible in the upper temperature range below 0°C, to vitrification. Unfortunately, a limit so far exists for human-sized organs at cryogenic temperatures. However, successful attempts exist to cool larger objects to temperatures as low as about -80°C. Transplantations of cooled and rewarmed organs have also taken place. Temperatures and methods vary widely in this range. No procedure generally applicable to human sized organs can yet be foreseen.

9.1 Cells

A large number of cell types – especially in cultures – can today be cryopreserved and stored in liquid nitrogen without major losses. Information on the required cooling rates, which are usually a few degrees per minute, is given by Mazur et al. (1972). Some special cell types should be mentioned here.

Stem cells from a wide variety of sources including those from neural tissue have been cryopreserved (Ballen et al 2013; Ballen et al 2009; Bojic et al 2014; Gluckman et al 1989; Harris 2014; Hilkens et al 2016; Huang et al 2019; Hunt 2017; Kashuba et al 2014; Kawata et al 2012; Ochiai et al 2021; Sun et al 2016; Weissman 2000; Xie et al 2022).

Heart muscle cells from chicks were frozen as early as 1968 by Schöpf and Ebner. Of these, only a very small percentage of cells capable of contraction survived. Rat heart cells were also frozen and kept at -180-190°C for up to three days. Normal appearance and normal motility were achieved. Human cardiac muscle cells have also been cryopreserved and were normal after thawing. Cells were also isolated from tissue that had already been cryopreserved and thawed, and could be maintained in culture. In addition, whole layers of muscle cells were vitrified for repair of cardiac muscle tissue. DMSO has a toxic effect on cardiac muscle cells at temperatures above zero (Alink GM et al. 1977; -1978; Bustamante, Jachimowicz 1988; Carmine et al. 2014; Hak et al. 1973; Kasten, Yip 1974; Ohkawara et al. 2018; Wollenberger 1967; -1967a).

Müller cells of the retina are glial cells, as the latter are found everywhere in the nervous system. They belong to the so-called macroglia. They play a role in the functioning of the retina, but also in pathological processes and they have been cryopreserved (Biedermann et al. 2002).

Today, sperm and oocytes can be vitrified with sufficient safety (Gosden 2011; Mazur, Seki 2011).

Sperm have first been frozen in the 1950s.and can be used in cases of impending fertility damage and artificial insemination. Here, we will only point out the importance of this method (Rodriguez-Wallberg et al. 2019). Slow freezing of sperm decreases motility (Mossad et al. 1994). Vitrification damages DNA less, maintains motility better, and requires less time and cost (Li et al. 2019; Riva et al. 2018; Vutyavanich et al. 2010). Sperm from dead Spanish red deer could be cryopreserved using different methods (Medina-Chavez et al. 2022). Special vitrification methods have been used for horse sperm (Devireddy et al. 2002; Oldenhof et al. 2017; Pruß et al. 2021). Long-term storage (up to 40 years) does not affect sperm fertilization ability (Feldschuh et al. 2005; Horne 2004; Szell et al. 2013).

Oocytes can be vitrified (Du et al. 2022) although they have special properties (Paynter et al. 1999). In calves, birth weight and liver size were increased when vitrified as mature oocytes (Jacobsen et al. 2000). The high-water content of oocytes makes them more susceptible to cooling damage than embryos and they achieve lower birth rates than embryos (Hudson et al. 2017). Oocytes are also discussed in the context of ovarian tissue (s. below).

> Background information
> Harmful effects of cryopreservation on oocytes (reviewed in Angarita et al. 2016) are found as hardening of the zona pellucida (Matson al. 1997), damage to the oocyte meiotic spindle, damage to the cytoskeleton, and granular cortical damage from ice crystals (Boiso et al. 2002).

Compared to vitrification lower fecundity rates (Pickering et al. 1991) and survival (Hochi et al. 2001; Jin 2014) were found with slow freezing (Edgar, Gook 2012).

Red blood cells are a readily available test material. However, their cryopreservation is somewhat delicate. When red blood cells are cryopreserved, the cryoprotectant must be enriched step by step to prevent damage due to fluctuations in osmotic pressure. If the damage is high, red blood cells will dissolve (hemolysis).

> Background information
> Using 70% VM-1 (the vitrifying antifreeze solution from Cryonics Institute), there was no immediate dissolution of sheep red blood cells when the solution (VM1) was added gradually at room temperature or near 0°C. Microscopic changes were also minor. However, when the solution was added in a single step, it caused the red blood cells to dissolve and more strongly even at low temperatures. VM1 contains ethylene glycol and dimethyl sulfoxide (DMSO). DMSO is a better glass former than ethylene glycol, but stepwise addition of a DMSO solution to 70% results in immediate, complete dissolution of the blood cells.

Unfortunately, it turns out that the test with red blood cells, although simple, is poor to evaluate mathematically and also too insensitive to detect finer damage (de Wolf, de Wolf 2013). They have recently been successfully cryopreserved (Murray et al. 2022).

9.2 Embryos

It has already been mentioned, not only cells but also entities made of many cells such as embryos can be cryopreserved and revived. Rall and Fahy vitrified mouse embryos in 1985 in a highly concentrated solution they called VS1. It allowed slow cooling, but viability was maintained only with rapid rewarming.

Embryo cryopreservation is now the gold standard for fertility preservation with high pregnancy rates (Angarita; et al. 2016; McLaren, Bates 2012).

Embryos from 2-cells and 4-cells have the best survival rates.

Vitrification has often been superior to slow freezing in terms of embryo survival rates and pregnancy (Keskintepe et al. 2009; Loutradi et al. 2008; Valojerdi et al. 2009).

For equine embryos, the conventional method with slow freezing and low doses of antifreeze showed no difference to vitrification for early developmental stages of equine embryos such as so-called morulae or blastocysts (Massip et al. 2001; Oberstein et al. 2001; Young et al. 1997).

The dependence on the object is sometimes inexplicable. In bovine embryos, for example, compact multicellular morulae and blastocysts survive cryopreservation and consequent development better than even much smaller precursors.

Once again, this shows that it are not always the smallest and simplest biological objects that are best suited for research.

In porcine embryos (at the morula or blastocyst stage), damage to so-called microfilaments was found during vitrification. These are microscopically small filament molecules that belong to the so-called skeleton of cells. An improvement could be achieved if the filament molecules were disassembled into individual building blocks beforehand (Dobrinsky et al. 2000). In any case, pig embryos also survive vitrification (Kobayashi et al. 1998).

> Background information
> However, Uechi and co-workers (1999) found a reduced viability of vitrified embryos. Also, a lower developmental rate was seen in mouse embryos that had been vitrified at the 2-cell stage. This was

shown by comparison with untreated embryos or with those frozen in the conventional way.

The hearts of chick embryos beat again after cooling to -196°C and thawing (Gonzales, Luyet 1950).

Overall, it is true that embryos can be vitrified today (Gosden 2011; Kawasaki et al. 2020, but see fish embryos (above).

9.2.1 Breakthrough in human reproduction

Fertility preservation is an important field of medicine that is growing rapidly. Since 1978, when the first birth after fertilization in vitro was reported, at least 8 million babies have been conceived this way. Cryopreservation techniques play an important role in this success by allowing long-term storage of gametes and embryos without decreasing their quality for later use. Sperm and oocytes have already been mentioned (Rodriguez-Wallberg et al. 2019).

In 1983, there was the first human pregnancy after cryopreservation (Trounson, Mohr) of a multicellular embryo cooled to the temperature of liquid nitrogen with gradual addition of DMSO. Time was given for the concentrations to equilibrate to avoid damage from osmotic pressure.

Since 1983, human embryos have been cryopreserved not only with DMSO but also with glycerol and propylene glycol. Since the first birth, more than half a million live births have been achieved with the help of embryo cryopreservation. Vitrification of the "germinal bubble" (blastocyst) is not an obstacle to normal development and birth in humans (the blastocyst is already composed of many cells forming a vesicle with a small inner marginal mound of cells the embryoblast, which later forms the organs of the embryo) (Bojic et al. 2021; El-Danasouri, Selmann 2001; Saito et al. 2000; Yokota et al. 2001).

Compared to natural births, there is no increased harm in children born after cryopreservation (Noyes et al. 2009).

Increasingly, women today are postponing childbearing to an older age with decreasing fertility. In addition to self-imposed postponement, there are medical situations that affect fertility, especially cancers where treatment interventions are germ-damaging. A number of other conditions also have the same negative effect (Baram et al. 2019; Condorelli, Demeestere 2019; Kieran, Shnorhavorian 2018; Singer et al. 2010; Zhao et al. 2019). Oocyte preservation is now particularly useful for women who want to plan a birth date independently (Cobo et al. 2016).

9.3 Organs or parts of organs

Mature tissues are built more complicated than very young embryos. It has already been mentioned and is worth mentioning that small tissue samples can nevertheless be frozen and resuscitated.

9.3.1 Tissue fractions from ovaries and testes

In addition to oocytes and embryonic tissue, tissue from ovaries is now also cryopreserved. (Courbiere et al 2006; Gook et al. 2021; Müller A 2012; Ting et al. 2012; -2013).

After thawing, the tissue can be transplanted back to the patient or immature oocytes can be isolated and grown to maturity in a tissue culture facility (McLaren, Bates 2012). In 2004, this resulted in the first birth after back-transplantation of ovarian tissue to a patient (Donnez et al. 2004). The advantage of reimplantation is the possibility of natural conception by the affected women (Beckmann et al. 2017; Meirow et al. 2016).

Cryopreservation of ovarian tissue is the only option for fertility preservation and normal conception for pre-adolescent girls and women who need immediate cancer treatment (Zhao et al. 2019). In this procedure, tissue is removed from the ovary containing oocytes and it is cryopreserved (Angarita et al. 2016; Campos 2011; Isachenko et al. 2012; Lotz et al. 2019). By reimplantation (In addition), hormone production by the ovarian tissue is also restored for several years and the onset of menopause is delayed (Anderson 2017; Biasin et al. 2015).

Cryopreservation of immature testicular tissue, on the other hand, is a method under development. Animal studies are promising. Healthy offspring have been conceived by transplantation of a frozen suspension of testicular cells or of tissue fragments. In humans, neither of these methods has been shown to be effective or safe (Wyns et al. 2010).

For boys before puberty who do not yet produce sperm, the only way to achieve and maintain fertility is to harvest testicular tissue — when they lose their testicles — and cryopreserve sperm stem cells. It is intended to restore reproductive capacity in boys after treatment with harmful side effects leading to sterility (Yang et al. 2019). The tissue is intended to be transplanted back into the testes later (e. g., after cancer therapy) to allow sperm to form. In trials, the tissue has been both frozen and vitrified (Brinster 2007; Zhao et al. 2019). However, carryover of cancer cells during transplantation has not yet been completely ruled out (Hudson et al. 2017) (see also Bojic et al. 2021 for fertility preservation).

9.3.2 Whole small organs

Ovaries from different animals — that is, whole organs — could produce offspring using a freeze-thaw method after vitrification and rewarming (Arav et al. 2005; Hasegawa et al. 2006). The largest ovaries cryopreserved were from sheep. Three ovaries were placed in liquid nitrogen and transplanted to sheep after thawing. After 6 years, 2 of them were normal and one was reduced in size (Arav et al. 2005; 2010).

9.3.3 Different sized organs, different temperatures from frosty to cryogenic

In smooth muscle, no ice crystals formed in a DMSO solution at -21°C. The ability of the muscle cells to contract was retained by 80% (Tailer and Pegg 1983).

The cornea of the eye (cornea) can be stored cooled for extended periods of time in unaltered form. In organ culture, corneas have been kept for 4 weeks (Armitage 2011). Corneal transplants are always needed. According to the WHO, nearly 15 million people require corneal surgery (Whitcher et al. 2001). There are far too few corneal donations (Golchet et al. 2000). It would be desirable to keep the corneas longer. For this purpose, corneas have also been vitrified with VS1, but the necessary high heating rates to avoid devitrification could not be achieved (Armitage 1991). The first successful cryopreservation of rabbit, cat, and human corneas was performed with 7.5% DMSO and 10% sucrose. Although the cryopreserved corneas were successfully transplanted, damage was found to the cells (endothelial cells) lining the cornea internally as a result of the freezing (Armitage 2002). In other experiments these cells were preserved, but the swelling of the cornea could not be controlled (Wusteman et al. 2008). Indeed, cells forming a simple layer are more sensitive to deep freezing than cells in a suspension. These cells have a very limited proliferation rate in the human eye and cells that are lost are not replaced by cell division (Sames 1980). Corneas from rabbits and humans have also been vitrified with 6.8 molar propane-1,2-diol and the function of these inner wall cells was preserved (Armitage et al. 2002; Armitage 2011). However, this technique requires very high solute concentrations and is time consuming.

> Background information
> Recently, a procedure using intermittent slow cooling (1°C/min) to a temperature of -35°C for porcine corneas and -45°C for human corneas prior to storage in liquid nitrogen was successfully used to preserve inner wall cells. A combination of 5% DMSO and 6% hydroxyethyl starch was used for this purpose (Marquez-Curtis et al. 2017).
> However, fresh corneas work better.
> The failure of cryopreservation in the inner cells could also be due to the fact that these cells pump water out of the cornea and are therefore likely to be very rich in water. A medium with increased concentration of solutes including those that are not cell-permeable might be the solution. In vivo cells are normally prevented from dividing by the lack of appropriate growth factors in the aqueous humor of the eye from which they feed (Chen et al. 1999). In contact with blood or serum, cultures of these cells grew well even when the culture solution was mixed with aqueous humor from eyes. We achieved 27 doublings of cell area in the cultures. Losses of cells are compensated by broadening neighbor cells when a dense layer exists. Thus, the number of cells in both the eye and the cultures decreases with increasing time. Colonization of the corneas with such cells is possible. This means that both the cells and the corneas can be cryopreserved and it should be possible to reunite them.

However, they might have to be taken from other species, as human cells of this type proliferate only slightly (Marquez-Curtis et al. space 2017; Sames 1980; Sames, Lindner 1982).

Overall, it is hoped that cryopreservation will allow trouble-free long storage of donor corneas in the near future.

Heart valves are also very thin structures. They were cooled using various methods. The survival of their connective tissue cells (fibroblasts) has been demonstrated. When cryopreserved, more heart valves survived transplantation for 10 years in the hearts of patients than if they had been kept at 4°C before insertion (see at Armitage 1991). Huber et al (2012) were able to cool heart valves in 10% DMSO to -80°C without ice formation. They also survived thawing.

Rat cardiac muscle cells survived in tissue samples for 1-24 weeks at -196°C. They were capable of beating and growing. The larger the pieces were, the fewer living cells were obtained after thawing. Such cells endured deep freezing in 7.5% DMSO (Alink et al. 1976ab; Alink et al. 1978; Yokomuro et al. 2003). Small samples of rat cardiac muscle have beaten again after cooling in liquid nitrogen and rewarming if they contained conduction tissue (Banker et al. 1991; Luyet 1969; -1969b;). In another experiment, sections of myocardial tissue from rats and humans were cooled to -80°C and rewarmed. It was found that the tissue temporarily showed some signs of life but did not recover. The methods remain to be verified (Kuhn 2013). After cryopreservation of cardiac muscle samples with DMSO, 70% respiration was preserved in the respiratory organelles, but the coupling of oxygen transfer and energy storage was impaired. The cryoprotectant could be responsible for the damage (Meyer et al. 2016).

Whole hearts are among the best studied organs, but have not been able to be cooled lower than -45°C.

Hearts above a size of that of the rabbit heart have not, to our knowledge, been cooled to cryogenic temperatures. Vitrification of whole hearts is also not known to us.

Rapatz (1970) studied the freezing and recovery of frog hearts, Leunissen, Piatnek-Leunissen (1968) the freezing of rat hearts within the body. Cooling of whole hearts below the freezing point (Karow Jr. et al. 1965) was done quite predominantly on hearts of small animals, since a general problem of cryonics is the very slow soaking of tissues from the surface when cooling by mere immersion. Many studies have been performed at temperatures not much below 0°C. Transplantation has also been tried. Butanediol, antifreeze proteins, and ethanol have been tested as antifreeze agents in various experiments. Hearts that were cooled below 0°C recovered better than those that were kept above O° (Amir et al. 2004; -2005; Banker et al. 1992; -1992a; Barsamian et al.; Elami et al. 2008; Fahy et al. 2004a; Kato et al. 2012; Offerjins 1971; Offerijns, Krijnen 1972; Offerjins, Ter Welle 1974; Sakaguchi et al. 1998; Wang et al.1992; Yang et al. 1993; Zhu et al. 1992).

Hearts perfused with various cryoprotective agents recovered from cooling to -20 to -40°C (Karow Jr. et al. 1965). However, in the rat heart perfused with DMSO, flow in the coronary arteries was impaired (Fahy, Karow 1977).

Hearts from 27-36 kg pigs — already larger hearts — were cooled to -3°C or 4°C and could be stored in a refrigerator that uses variable magnetic fields to prevent crystal formation by oscillating water molecules. It cools all parts of the tissue simultaneously and therefore can be used for thick heart walls. It also reduces the amount of harmful oxygen products, no antifreeze is needed, and it could be used for years of storage. However, no attempts have been made to make the hearts beat (Kato et al. 2012; Seguchi et al. 2015).

In blood vessels, as in the cornea, the maintenance of the cells that paper the inside (endothelial cells) is a major problem. Metabolic performance, ability to contract in response to nerve impulses, and appearance were ways of assessing the maintenance of the vessels as a whole. Veins and coronary arteries were successfully cryopreserved by the freeze-thaw method (see in Fuller, Grout 1991, Muller-Schweinitzer 1988; -et al. 1997). Carotid artery pieces up to 25mm in length from rabbits were vitrified with VS55 and rewarmed viable. This was possible by adjusting the heating and cooling rates and cooling in nitrogen vapor to -130°C. (Baicu et al. 2006). In canine coronary arteries, vasodilatory responses were found after cryopreservation and thawing even when cooled to -196°C (Müller-Schweinitzer et al. 1997).

Pancreatic islets are the hormone glands within the pancreas that produce insulin for the metabolism of sugar. They can now be isolated and allowed to survive in culture. There was a successful transplantation of islets frozen by the freeze-thaw method onto diabetic rats. Blood glucose normalized thereafter. However, more of the cryopreserved islets were needed compared to freshly harvested islets. Islets from dogs were also tested in the same way after successful cryopreservation. Vitrification of the islets is also possible. However, vitrification has been shown to offer no advantage over conventional cooling, which in this case was actually highly superior (Fuller, Grout 1991; Langer et al. 1999). On the other hand, Mukherjee et al. (2005) found vitrification to be the best form of cryopreservation of encapsulated pancreatic cells (reviewed in Marquez-Curtis 2022). Recently, the state of the art of research was summarized and a positive conclusion of vitrification was drawn in terms of survival, function, and application (Bischof and Finger 2022; Kojayan et al 2018).

Kidneys have been studied in a variety of ways. Marco-Jimenez et al. (2015) have vitrified embryonic kidney precursors as a source of grafts, as well as early embryonic tissue (primordial tissue) from which kidney corpuscles can develop after thawing and transplantation (Garcia-Dominguez et al. 2016).

A rabbit kidney has a volume of about 10ml, weighs about 7-8g, and following vitrification can be brought to the temperature of liquid nitrogen within 2 days without freeze fractures (as we recall, not even 5 days is sufficient for slow cooling in whole patients to avoid freeze fractures) (Fahy et al. 1990; -1990a).

Like hearts, kidneys were difficult to cool to temperatures below -45°C and only occasionally cooled to -80° (Fahy et al. 2004). A rabbit kidney was cooled to -130°C. A special conductive heating technique combined with perfusion resulted in successful heating (Fahy et al. 2009; cf. Wowk, Corral 2013). This kidney from a rabbit had been successfully perfused with 7.5 M M22 as antifreeze. Alone It completely supplied a living animal after thawing (Fahy, Ali 1997; Fahy et al. 2009; -2013; Kheirabadi, Fahy 2000). However, it was only one kidney that survived in this manner among several ones. It was not cooled to liquid nitrogen temperature to avoid cracking and flow- and electrolyte disturbances occurred. In previous experiments, the same was achieved after cooling to -45°C and further preliminary experiments already indicated survival after vitrification (Fahy et al. 2004; -2004a; -2006), so this success was not accidental. Although it has not yet been repeated and did not work for all kidneys in this experiment, it is clear evidence of the possibility of cryopreservation of organs of this size, which are as intricately built as organs of larger mammals. Later Han et al. (2023) developed a safe and reproducible method to cryopreserve, reanimate and transplant rat kidneys. They used perfusion and vitrification, cooling to -150°C followed after storage for 100 days by nanowarming, a method which in principle should function also with larger kidneys.

Whole hind legs of rats were cooled to -140°C by directional cooling (Bahari et al. 2018) before being placed into liquid nitrogen. They were then successfully resuscitated. While these legs could not function because the nerve cords for movement were cut during amputation, the rest of the legs were alive (Arav et al. 2017). Rat legs are small but complexly made up of multiple organs, so this represents considerable progress.

Amputated human fingers have been cryopreserved and stored in liquid nitrogen for up to 81 days and transplanted. They regained and maintained function (Wang et al. 2019; Wang et al. 2020).

The following experiments used different methods and temperatures. They were only partially successfully repeated and individual organs showed damage due to ice formation. However, all organs listed here survived.

Uteruses of pigs contracted several times after cooling to -130°C and subsequent heating (Dittrich et al. 2006).

Dog intestine was cooled to the temperature of liquid nitrogen. It was severely damaged but capable of self-repair. An exception was the blood vessels, but they remained open (Hamilton et al. 1973).

The canine spleen survived chilling to freezing temperatures and transplantation (Barner, Scheck 1966).

With the canine ureter, cryopreservation and transplantation succeeded (Barner et al. 1963).

Dog lungs also survived temperatures below 0°C (Okaniwa et al. 1973).

Liver transplants are in demand in medicine because the detoxification function of the liver is essential for survival. The problem of preserving livers in a living state for the long term has been addressed many times.

Parts of the liver (cell clots or microbeads) and encapsulated liver cells can temporarily take over liver function and thus save life. They can be cryopreserved (Jitraruch et al. 2017; Massie et al. 2011).

Liver cells can be protected by a special form of glucose, which is not broken down in metabolism and does not participate in metabolism at all. It goes into the cells and prevents shrinkage (Sugimachi et al. 2006).

Meanwhile, rat livers were successfully transplanted after 4 days of hypothermia using a method that spares the cells of the inner lining of vessels (the liver contains an extensive specialized network of blood vessels). Polyethylene glycol with various functions for cell protection has been used (Berendsen et al. 2014; Elliot, Wang. 2017).

Livers were the largest organs that could be frozen, although not at cryogenic temperatures. Rat livers were cooled to -3 to 4°C for 6-24 hours after which bile production and microscopic structure were preserved (Rubinsky et al. 1994; Soltys et al. 2001). Livers were maintained for 77-96 hours even at -6°C with PEG and 3-O-methylglucose. Organ survival was 100% and better than at 4°C. However, after 96 hours, survival rate and ATP content were decreased. Rat livers and canine livers were studied in different but similar studies with similar results (Bruinsma et al. 2014; Bruinsma, Uygun 2017).

A pig liver was kept at -20°C for half an hour and then heated. It was implanted into another pig. It was demonstrated that the liver functioned in the new host. After killing the pig, the liver was removed and the cells were examined with the result that they were alive. The directional freezing method was used. That it was alive was demonstrated by the flow of blood and metabolic activity (Gavish et al. 2008). Other livers also regained partial function even after cooling to -60°C. The pancreas was cooled by the same method and with the same results (Zimmermann et al. 1971).

Bone tissue, in terms of cell viability and biochemical assessment, shows no different behavior after cryopreservation than when fresh and used as graft material (Nevo et al. 1983).

Cartilage tissue contains a large percentage of proteoglycans with antifreeze activity around the cells. Therefore, its survival at low temperatures is of particular interest. Its successful cryopreservation has already been discussed above in connection with the antifreeze effect of proteoglycans (see also Kiefer et al. 1989). However, in intervertebral discs that had been frozen in DMSO, proteoglycan formation by the cells was dramatically reduced (Matsuzaki et al 1996).

Teeth that have not fully matured can be transplanted but are difficult to keep alive in an isolated state for a long time. This is problematic for stockpiling, but also when a tooth graft must be preserved until the wound of an extracted tooth has healed. According to one case report, a tooth was cooled in a programmed freezer with a vibrating magnetic field that prevents ice formation. This was followed by storage at -150°C and successful transplantation, demonstrating life preservation of the tooth and healing (Kaku M et al. 2015).

There is no doubt that the successes discussed here make us optimistic. There remain problems with cooling of large organs to cryogenic temperatures, cracking and damaging effects of cryoprotectants.

Literature

Alink GM et al (1976a) The effect of cooling rate and of dimethyl sulfoxide concentration on low temperature preservation of neonatal rat heart cells. Cryobiology 13:295-304

Alink GM et al (1976b) The effect of cooling rate and of dimethyl sulfoxide concentration on the ultrastructure of neonatal rat heart cells after freezing and thawing. Cryobiology 13:305-316

Alink GM et al (1977) Three-step cooling: a preservation method for adult rat heart cells. Cryobiology 14:409-417

Alink GM et al (1978) Viability and morphology of rat heart cells after freezing and thawing of the whole heart. Cryobiology 15:44-58

Amir G et al (2004) Subzero nonfreezing cryopresevation of rat hearts using antifreeze protein I and antifreeze protein III. Cryobiology 48:273-282

Amir G et al (2005) Improved viability and reduced apoptosis in sub-zero 21-hour preservation of transplanted rat hearts using anti-freeze proteins. J Heart Lung Transplant 24:1915-1929

Anderson RA (2017) Ovarian tissue cryopreservation for fertility preservation: clinical and research perspectives. Hum Reprod Open 2017(1): hox001

Angarita AM et al (2016) Fertility preservation: a key survivorship issue for young women with cancer. Front Oncol 6:102

Arav A et al (2005) Oocyte recovery, embryo development and ovarian sheep ovary. Hum Reprod 20:3554-3559

Arav A et al (2010) Ovarian function 6 years after cryopreservation and transplantation of whole sheep ovaries. Reprod Biomed Online 20:48-52

Arav A et al (2017) Rat Hindlimb Cryopreservation and Transplantation: A step toward "organ banking". Am J Transplant 17:2820-2828

Armitage WJ (1991) Preservation of viable tissues for transplantation. In Fuller BJ, Grout BWW (eds) Clinical Applications of Cryobiology. CRC Press, Boca Raton, pp 170-189

Armitage WJ (2002) Recovery of endothelial function after vitrification of cornea at -110 degrees C. Invest Ophthalmol Vis Sci 43:2160-2164

Armitage WJ (2011) Preservation of human cornea. Transfus Med Hemother 38:143-147

Bahari L et al (2018) Directional freezing for the cryopreservation of adherent mammalian cells on a substrate. PloS one 13:e0192265

Baicu S et al (2006) Vitrification of carotid artery segments: An integrated study of thermophysical events and functional recovery toward scale-up for clinical application. Cell Preserv Technol 4:236-244

Ballen KK et al (2009) Umbilical cord blood donation: public or private? Bone Marrow Transplant 50 (2015) 1271-1278

Ballen KK et al (2013) Umbilical cord blood transplantation: the first 25 years and beyond. Blood 122:491-498

Banker M et al (1991) Freezing preservation of the mammalian cardiac explant IV. Functional recovery after 8-hour freezing. Curr Surg 48:428-430

Banker MC et al (1992) Freezing preservation of the mammalian cardiac explant. II. comparing the protective effect of glycerol and polyethylene glycol. Cryobiology 29:87-94

Banker MC et al (1992a) Freezing preservation of the mammalian heart explant. III. tissue dehydration and cryoprotection by polyethylene glycol. J Heart Lung Transplant 11:619-623

Baram S et al (2019) Fertility preservation for transgender adolescents and young adults: a systematic review. Hum Reprod Update 25:694-716

Barner HB et al (1963) Survival of canine ureter after freezing. Surgery 53:344-347

Barner HB, Scheck E (1966) Autotransplantation of the frozen-thawed spleen. Arch Pathol Lab Med 82:267-271

Barsamian EM et al (1960) Preliminary studies on the transplantation of supercooled hearts. Plast Reconstr Surg Transplant Bull 25:405-406

Beckmann MW et al (2017) Operative techniques and complications of extraction and transplantation of ovarian tissue: the Erlangen experience. Arch Gynecol Obstet 295:1033-1039

Berendsen TA et al (2014) Supercooling enables long-term transplant survival following 4 days of liver preservation. Nat Med 20:790-793

Biasin E et al (2015) Ovarian tissue cryopreservation in girls undergoing haematopoietic stem cell transplantation: experience of a single centre. Bone Marrow Transplant 50:1206-1211

Biedermann B et al (2002) Patch-clamp recording of Muller glial cells after cryopreservation. J Neurosci Methods 120:173-178

Bishop JC, Finger EB (2022) Cryopreservation of Pancreatic Islets Experimental Data Repository 2022. Retrieved from the Data Repository for the University of Minnesota, https://doi.org/10.13020/yrva-zr31

Boiso I et al (2002) A confocal microscopy analysis of the spindle and chromosome configurations of human oocytes cryopreserved at the germinal vesicle and metaphase II stage. Hum Reprod 17:1885-1891

Bojic S et al (2014) Dental stem cells—characteristics and potential. Histol Histopathol 29:699-706

Bojic S et al (2021) Winter is coming: the future of cryopreservation. BMC Biol 19, 56

Brinster R (2007) Male germline stem cells: from mice to men. Science 316:404-405

Bruinsma BG, Uygun K (2017) Subzero organ preservation; the dawn of a new ice age? Curr Opin Organ Transplant 22:281-286

Bruinsma BG et al (2014) Subnormothermic machine perfusion for ex vivo preservation and recovery of the human liver for transplantation. Am J Transplant 14:1400-1409

Bustamante JO, Jachimowicz D (1988) Cryopreservation of human heart cells. Cryobiology 25:394-408

Campos JR et al (2011) Cryopreservation and fertility: current and prospective possibilities for female cancer patients. ISRN Obstet Gynecol (2011) ID 358113

Carmine G et al (2014) A novel method for isolating and culturing human cardiomyocytes from cryopreserved tissues. Biophys J 106, 564a

Chen KH et al (1999) TGF-beta 2 in aqueous humor suppresses S-phase entry in cultured corneal endothelial cells. Invest Ophthalmol Vis Sci 40:2513-2519

Cobo A et al (2016) Oocyte vitrification as an efficient option for elective fertility preservation. Fertil Steril 105:755-764

Condorelli M, Demeestere I (2019) Challenges of fertility preservation in non-oncological diseases. Acta Obstet Gynecol Scand 98:638-646

Courbiere B et al (2006) Cryopreservation of the ovary by vitrification as an alternative to slow-cooling protocols. Fertil Steril 86:1243-1251

Devireddy RV et al (2002) Cryopreservation of equine sperm: optimal cooling rates in the presence and absence of cryoprotective agents determined using differential scanning calorimetry. Biol Reprod 66:222-231

De Wolf A, de Wolf C (2013) Human cryopreservation research at advanced neural biosciences. In Sames KH (ed) Applied Human Cryobiology, vol. 1, ibidem, Stuttgart, pp 45-59

Dittrich R et al (2006) Successful uterine cryopreservation in an animal model. J Hormone Metab Res 38:141-145

Dobrinsky JR et al (2000) Birth of piglets after transfer of embryos cryopreserved by cytoskeletal stabilization and vitrification. Biol Reprod 62:564-570

Donnez J et al (2004) Livebirth after orthotopic transplantation of cryopreserved ovarian tissue. Lancet 364:1405-1410

Du X et al (2022) Artificially Increasing Cortical Tension Improves Mouse Oocytes Development by Attenuating Meiotic Defects During Vitrification. Front Cell Dev Biol 10:876259. doi: 10.3389/fcell.2022.876259

Edgar DH, Gook DA (2012) A critical appraisal of cryopreservation (slow cooling versus vitrification) of human oocytes and embryos. Hum Reprod Update 18:536-554

Elami A et al (2008) Successful restoration of function of frozen and thawed isolated rat hearts. Thorac Cardiovas Surg 135:666-672

El-Danasouri I, Selman H (2001) Successful pregnancies and deliveries after a simple vitrification protocol for day 3 human embryos. Fertile Sterile 76:400-402

Elliott GD, Wang S (2017) Cryoprotectants: a review of the actions and applications of cryoprotective solutes that modulate cell recovery from ultra-low temperatures. Cryobiology 76:74-91

Fahy GM, Ali SE (1997) Cryopreservation of the mammalian kidney. II Demonstration of immediate ex vivo function after introduction and removal of 7.5 M cryoprotectant. Cryobiology 35:114-131

Fahy GM, Karow AM Jr (1977) Ultrastructure-function correlative studies for cardiac cryopreservation. V. Absence of a correlation between electrolyte toxicity and cryoinjury in the slowly frozen, cryoprotected rat heart. Cryobiology 14:418-427

Fahy GM et al (1990) Cryoprotectant Toxicity and Cryoprotectant Toxicity Reductio. In Search of Molecular Mechanisms. Cryobiology 27:247-268

Fahy GM et al (1990a) Physical problems with the vitrification of large biological systems. Cryobiology 27:492-510

Fahy GM et al (2004) Improved vitrification solution based on the predictability of vitrification solution toxicity. Cryobiology 42:22-35

Fahy et al (2004a) Cryopreservation of organs by vitrification: perspectives and recent advances. Cryobiology 48:157-178

Fahy GM et al (2006) Cryopreservation of complex systems: the missing link in the regenerative medicine supply chain. Rejuvenation Res. 9:279-291

Fahy GM et al (2009) Physical and Biological Aspects of Renal Vitrification. Organogenesis 5:167-175

Fahy GM et al (2013) Cryopreservation of precision-cut tissue slices. Xenobiotica 43:113-132

Feldschuh J et al (2005) Successful sperm storage for 28 years. Fertil Steril 84:P1017.e3-1017.e4

Fuller BJ, Grout BWW (eds) (1991) Clinical Applications of Cryobiology. CRC Press, Boca Raton

Garcia-Dominguez X et al (2016) First steps toward organ banks: vitrification of renal primordial. Cryo Letters 37:347-352

Gavish Z et al (2008) Cryopreservation of whole murine and porcine livers. Rejuvenation Res 11:765-772

Gluckman E et al (1989) Hematopoietic reconstitution in a patient with Fanconi's anemia by means of umbilical-cord blood from an HLA-identical sibling. New Engl J Med 321:1174-1178

Golchet G et al (2000) Why don't we have enough corneal donors? A literature review and survey. Optometry (St Louis, Mo) 71:318-328

Gonzales F, Luyet B (1950) Resumption of heart-beat in chick embryos frozen in liquid nitrogen. Biodynamica 7:1-5

Gook D et al (2021) Experience with transplantation of human cryopreserved ovarian tissue to a sub-peritoneal abdominal site. Hum Reprod Hum Reprod 36:2473-2483

Gosden R (2011) Cryopreservation: a cold look at technology for fertility preservation. Fertil Steril 96:264-268

Hak AM et al (1973) Toxic effects of DMSO on cultured beating heart cells at temperatures above zero. Cryobiology 10:244-250

Hamilton R et al (1973) Successful preservation of canine small intestine by freezing. J Surg Res 14:313-318

Han Z et al (2023) Vitrification and nanowarming enable long-term organ cryopreservation and life-sustaining kidney transplantation in a rat model. Nat Commun 14 3407 (art. Nr.)

Harris DT (2014) Stem cell banking for regenerative and personalized medicine. Biomedicines 2:50-79

Hasegawa A et al (2006) Pup birth from mouse oocytes in preantral follicles derived from vitrified and warmed ovaries followed by in vitro growth, in vitro maturation, and in vitro fertilization. Fertil Steril 86 (4 Suppl):1182-1192

Hilkens P (2016) Cryopreservation and banking of dental stem cells. Adv Exp Med Biol 951:199-235

Hochi S et al (2001) Effects of cooling and warming rates during vitrification on fertilization of in vitro-matured bovine oocytes. Cryobiology 42:69-73

Horne G (2004) Live birth with sperm cryopreserved for 21 years prior to cancer treatment: case report. Hum Reprod 19:1448-1449

Huang CY et al (2019) Human iPSC banking: barriers and opportunities. J Biomed Sci 26:87

Huber A et al (2012) Development of a simplified ice-free cryopreservation method for heart valves employing V space S83, and 83% cryoprotectant formulation. Biopreserv Biobank 10:479-484

Hudson JN et al (2017) New Promising Strategies in Oncofertility. Expert Rev Qual Life Cancer Care 2:67-68

Hunt CJ (2017) Cryopreservation: vitrification and controlled rate cooling. Methods Mol Biol 1590:41-77

Isachenko V et al (2012) Cryopreservation of ovarian tissue: detailed description of methods for transport, freezing and thawing. Gynecology 72:927-932

Jacobsen H et al (2000) Body dimensions and birth and organ weights of calves derived from in vitro produced embryos cultured with or without serum and oviduct epithelium cells. Theriogenology 53:17 61-1769

Jin B et al (2014) Survivals of mouse oocytes approach 100% after vitrification in 3-fold diluted media and ultra-rapid warming by an IR laser pulse. Cryobiology 68:419-430

Jitraruch S et al (2017) Cryopreservation of Hepatocyte Microbeads for Clinical Transplantation. Cell Transplant 26:1341-1354

Kaku M et al (2015) A case of tooth autotransplantation after long-term cryopreservation using a programmed freezer with a magnetic field. Angle Orthod 85:518-524

Karow AM Jr et al (1965) Preservation of hearts by freezing. Arch Surg. 91:572-574

Kashuba CM et al (2014) Rationally optimized cryopreservation of multiple mouse embryonic stem cell lines: I--Comparative fundamental cryobiology of multiple mouse embryonic stem cell lines and the implications for embryonic stem cell cryopreservation protocols. Cryobiology 68:166-175

Kasten FH, Yip DK (1974) Reanimation of cultured mammalian myocardial cells during multiple cycles of trypsinization-freezing-thawing. In Vitro 9:246-252

Kato H et al (2012) Subzero 24-hour nonfreezing rat heart preservation: a novel preservation method in a variable magnetic field. Transplantation 94:473-477

Kawasaki Y et al (2020) Carboxylated epsilon-poly-L-lysine, a cryoprotective agent, is an effective partner of ethylene glycol for the vitrification of embryos at various preimplantation stages. Cryobiology 97:245-249

Kawata et al (2012) Effects of DMSO (dimethyl sulfoxide) free cryopreservation with program freezing using a magnetic field on periodontal ligament cells and dental pulp tissues. Biomed Res 23:438-443

Keskintepe L et al (2009) Vitrification of human embryos subjected to blastomere biopsy for pre-implantation genetic screening produces higher survival and pregnancy rates than slow freezing. J Assist Reprod Genet 26:629-635

Kheirabadi BS, Fahy GM (2000) Permanent life support by kidneys perfused with a vitrifiable (7.5 molar) cryoprotectant solution. Transplantation 70:51-55

Kiefer GN et al (1989) The effect of cryopreservation on the biomechanical behavior of bovine articular cartilage. J Orthop Res 7:494-501

Kieran K, Shnorhavorian M (2018) Fertility Issues in Pediatric Urology. Urol Clin North Am 45:587-599

Kobayashi S et al (1998) Piglets produced by transfer of vitrified porcine embryos after stepwise dilution of cryoprotectants. Cryobiology 36:20-31

Kojayan GG et al (2018) Systematic review of islet cryopreservation. Islets 10:40-49

Kuhn SA (2013) Einfluss der Kryokonservierung auf die strukturelle Integrität und die Elektrophysiologie von Herzdünnschnitten. Ein innovatives Testsystem für Herzmedikamente. Dissertation, Universität Lübeck

Langer S et al (1999) Viability and recovery of frozen-thawed human islets and in vivo quality control by xenotransplantation. J Mol Med 77:172-174

Leunissen RL, Piatnek-Leunissen DA (1968) A device facilitating in situ freezing of rat heart with modified Wollenberger tongs. J Appl Physiol. 25:769-771

Li YX et al (2019) Vitrification and conventional freezing methods in sperm cryopreservation: A systematic review and meta-analysis. Eur J Obstet Gynecol Reprod Biol 233:84-92

Lotz L et al (2019) Ovarian tissue transplantation: experience from Germany and worldwide efficacy. Clin Med Insights Reprod Health. 13:1179558119867357. eCollection 2019

Loutradi KE et al (2008) Cryopreservation of human embryos by vitrification or slow freezing: a systematic review and meta-analysis. Fertil Steril 90:186-193

Lyet B (1969) Resumption of contractions after freezing in liquid nitrogen in small pieces of rat heart containing pacemaking centers. Cryobiology 6:246-248

Luyet B (1969b) Resumption of activity in spontaneously and in non-spontaneously beating pieces of frog's hearts after freezing in liquid nitrogen. Biodynamica 10:261-275

Marco-Jimenez F et al (2015) Vitrification of kidney precursors as a new source for organ transplantation. Cryobiology 70:278-282

Marquez-Curtis LA et al (2017) Expansion and cryopreservation of porcine and human corneal endothelial cells. Cryobiology 77:1-13

Marquez-Curtis LA et al (2022) Cryopreservation and post-thaw characterization of dissociated human islet cells. PLoS ONE 17:e0263005, https://doi.org/10.1371/journal.pone.0263

Massie I et al (2011) Cryopreservation of encapsulated liver spheroids using a cryogen-free cooler: high functional recovery using a multi-step cooling profile. Cryo Letters 32:158-165

Massip A (2001) Cryopreservation of embryos of farm animals. Reprod Domest Anim 36:49-55

Matson PL et al (1997) Cryopreservation of oocytes and embryos: use of a mouse model to investigate effects upon zona hardness and formulate treatment strategies in an in vitro fertilization program. Hum Reprod 12:1550-1553

Mazur P, Seki S (2011) Survival of mouse oocytes being cooled in a vitrification solution to -196° at 95° to 70,000°C/min and warmed at 610° to 118,000°C/min: A new paradigm for cryopreservation by vitrification. Cryobiology 62:1-7

Mazur P et al (1972) A two-factor hypothesis of freezing injury. Exper Cell Res 1:345-355

McLaren JF, Bates GW (2012) Fertility preservation in women of reproductive age with cancer. Am J Obstet Gynecol 207:455-462

Medina-Chávez DA et al (2022) Freezing Protocol Optimization for Iberian Red Deer (Cervus elaphus hispanicus) Epididymal Sperm under Field Conditions. Animals (Basel) 12:869, doi: 10.3390/ani12070869

Meirow D et al (2016) Transplantations of frozen-thawed ovarian tissue demonstrate high reproductive performance and the need to revise restrictive criteria. Fertil Steril 106:467-474

Meyer A et al (2016) Cardiac mitochondrial oxidative capacity is partly preserved after cryopreservation with dimethyl sulfoxide. Cryo Letters 37:110-114

Mossad H et al (1994) Impact of cryopreservation on spermatozoa from infertile men: implications for artificial insemination. Arch Androl 33:51-57

Mukherjee N et al (2005) Effects of cryopreservation on cell viability and insulin secretion in a model tissue-engineered pancreatic substitute (TEPS). Cell Transplant 14:449-456

Müller A et al (2012) Retransplantation of cryopreserved ovarian tissue: first live birth in Germany. Dtsch Arztebl Int 109:8-13

Müller-Schweinitzer E (1988) Cryopreservation of isolated blood vessels. Folia Haematol Int Mag Klin Morphol Blutforsch 115:405-409

Müller-Schweinitzer E et al (1997) Functional recovery of human mesenteric and coronary arteries after cryopreservation at -196 degrees C in a serum-free medium. J Vasc Surg 25:743-745

Murray A et al (2022) Red Blood Cell Cryopreservation with Minimal Post-Thaw Lysis Enabled by a Synergistic Combination of a Cryoprotecting Polyampholyte with DMSO/Trehalose. Biomacromolecules 23; 467-477

Nevo Z et al (1983) Fresh and cryopreserved fetal bones replacing massive bone loss in rats. Calcif Tissue Int 35:62-69

Noyes N et al (2009) Over 900 oocyte cryopreservation babies born with no apparent increase in congenital anomalies. Reprod Biomed Online 18:769-776

Oberstein N et al (2001) Cryopreservation of equine embryos by open pulled straw, cryoloop, or conventional slow cooling methods. Theriogenology 55:607-613

Ochiai J et al (2021) Development of multilayer mesenchymal stem cell sheets. Int. J. Transl. Med 1:4-24

Offerijns FG (1971) Experiments on long-term preservation of rat heart. Br Heart J 33:149

Offerijns FG, Krijnen HW (1972) The preservation of the rat heart in the frozen state. Cryobiology 9:289-295

Offerijns FG, Ter Welle HF (1974) The effect of freezing, of supercooling and of DMSO on the function of mitochondria and on the contractility of the rat heart. Cryobiology 11:152-159

Ohkawara H et al (2018) Development of a vitrification method for preserving human myoblast cell sheets for myocardial regeneration therapy. BMC Biotechnol 18:Article 56

Okaniwa G et al (1973) Studies on the preservation of canine lung at sub-zero temperatures. J Thorac Cardiovasc Surg 65:180-186

Oldenhof H et al (2017) Stallion sperm cryopreservation using various permeating agents: interplay between concentration and cooling rate. Biopres Biobank 15:422-431

Paynter SJ et al (1999) Temperature dependence of Kedem-Katchalsky membrane transport coefficients for mature mouse oocytes in the presence of ethylene glycol. Cryobiology 39:169-176

Pickering SJ et al (1991) Cryoprotection of human oocytes: inappropriate exposure to DMSO reduces fertilization rates. Hum Reprod 6:142-143

Pruß D et al (2021) High-throughput droplet vitrification of stallion sperm using permeating cryoprotective agent. Cryobiology 101:67-77

Rapatz G (1970) Some problems associated with the freezing of hearts. Cryobiology 7:157-162

Riva NS et al (2018) Comparative analysis between slow freezing and ultra-rapid freezing for human sperm cryopreservation. JBRA Assist Reprod 22:331-333

Rodriguez-Wallberg KA et al (2019) Ice age: cryopreservation in assisted reproduction — An update. Reprod Biol 19:119-126

Rubinsky B et al (1994) Freezing mammalian livers with glycerol and antifreeze proteins. Biochem Biophys Res Commun 200:732-741

Saito H et al (2000) Application of vitrification to human embryo freezing. Gynecol Obstet Invest 49:145-149

Sakaguchi H et al (1998) Subzero non freezing storage (1 degree C) of the heart with University of Wisconsin solution and 2-3 butanediol. Transplant Proc 30:58-59

Sames K (1980) Morphologische und histochemische Untersuchungen über das in-vitro und in-vivo Altern von Corneaendothel und Trabeculum corneosclerale. Universität Erlangen, Habilitationsschrift

Sames K, Lindner J (1982) Changes in cell cultures of bovine corneal endothelium cells as related to donor age and number of passages in vitro. Akt Gerontol 12:206-212

Schöpf-Ebner et al (1968) Pulsatile activity of isolated heart muscle cells after freezing storage. Cryobiology 4:200-203

Seguchi R et al (2015) Subzero 12-hour nonfreezing cryopreservation of porcine heart in a variable magnetic field. Transplant Direct 1:9 e33

Singer ST et al (2010) Fertility potential in thalassemia major women: current findings and future diagnostic tools. Ann NY Acad Sci 1202:226-230

Soltys KA et al (2001) Successful nonfreezing, subzero preservation of rat liver with 2,3-butanediol and type I antifreeze protein. J Surg Res 96:30-34

Sugimachi K et al (2006) Nonmetabolizable glucose compounds impart cryotolerance to primary rat hepatocytes. Tissue Eng 12: 579-588

Sun C et al (2016) Fundamental principles of stem cell banking. Adv Exper Med Biol 95:31-45

Szell AZ et al (2013) Live births from frozen human semen stored for 40 years. J Assist Reprod Genet (JARG) 30:743-744

Taylor MJ, Pegg DE (1983) The effect of ice formation on the function of smooth muscle tissue stored at -21 or -60°C. Cryobiology 20:36-40

Ting AY et al (2012) Synthetic polymers improve vitrification outcomes of macaque ovarian tissue as assessed by histological integrity and the in vitro development of secondary follicles. Cryobiology 65:1-11

Ting AY et al (2013) Morphological and functional preservation of pre-antral follicles after vitrification of macaque ovarian tissue in a closed system. Hum Reprod 28:1267-1279

Trounson A, Mohr L (1983) Human pregnancy following cryopreservation, thawing and transfer of an eight-cell embryo. Nature 305:707-709

Uechi H et al (1999) Comparison of the effects of controlled-rate cryopreservation and vitrification on 2-cell mouse embryos and their subsequent development. Hum Reprod 14:2827-2832

Valojerdi MR et al (2009) Vitrification versus slow freezing gives excellent survival, postwarming embryo morphology and pregnancy outcomes for human cleaved embryos. J Assist Reprod Genet 26:347-354

Vutyavanich T et al (2010) Rapid freezing versus slow programmable freezing of human spermatozoa. Fertil Steril 93:1921-1928

Wang T et al (1992) Freezing preservation of the mammalian cardiac explant. V. Cryoprotection by ethanol. Cryobiology 29:470-477

Wang ZT et al (2019) Replantation of cryopreserved fingers: an "organ banking" breakthrough. Plast Reconstr Surg 144:679–683

Wang J et al (2020) Cryopreservation and transplantation of amputated finger. Cryobiology 92:235-240

Weissman IL (2000) Stem cells: units of development, units of regeneration, and units in evolution. Cell 100:157-168

Whitcher JP et al (2001) Corneal blindness: a global perspective. Bull World Health Organ 79:214-221

Wollenberger A (1967) Survival of potentially beating single heart cells in the frozen state. In Tanz RD et al (eds) Factors Influencing Myocardial Contractility. Academic Press, New York, pp 317-327

Wollenberger A et al (1967a) Cultivation of beating heart cells from frozen heart cell suspensions. Natural Sciences 54:174

Wowk B, Corral A (2013) 023 Adaptation of a commercial diathermy machine for radiofrequency warming of vitrified organs. Cryobiology 67:404

Wustemann MC et al (2008) Vitrification of rabbit tissues with propylene glycol and trehalose. Cryobiology 56:62-71

Wyns C et al (2010) Options for fertility preservation in prepubertal boys. Hum Reprod Update 16:312-328

Yang H et al (2019) Non-oncologic indications for male fertility preservation. Curr Urol Rep 20:51

Yang X et al (1993) Subzero nonfreezing storage of the mammalian cardiac explant. I. Methanol, ethanol, ethylene glycol, and propylene glycol as colligative cryoprotectants. Cryobiology 30:366-375

Yokomuro H et al (2003) Optimal conditions for heart cell cryopreservation for transplantation. Mol Cell Biochem 242:109-114

Yokota Y et al (2001) Birth of a healthy baby following vitrification of human blastocysts. Fertil Steril 75:1027-1029o

Young CA et al (1997) Cryopreservation procedures for day 7-8 equine embryos. Equine Vet J Suppl 25:98-102

Xie J et al (2022) Principles and protocols for post-cryopreservation quality evaluation of stem cells in novel biomedicine. Front. Pharmacol. 13:907943, doi: 10.3389/fphar.2022.907943

10 Important and promising starting points in the laboratory as well as in nature

Summary

You don't have to give up a brain after 9 minutes of cardiac arrest. It at this time is not yet really dead.

For the preservation of artificial tissues and organs, one should rely on methods of cryopreservation as developed in cryonics.

Even cryonics devotees are only incompletely aware that mammals have already been cooled to temperatures (just) below 0°C in the laboratory and resuscitated, and that this is also possible in nature. In many cold-blooded animals, this is commonly the case. Cryogenic temperatures rarely occur in nature. However, they are practically standard for our preservation programs. Therefore, it is interesting that some small multicellular animals can be brought to such temperatures in the laboratory and survive them.

10.1 Our irreplaceable biocomputer, the brain and nervous system

The brain is an organ that is individually characterized and irreplaceable. It is a part of the nervous system whose tissues and organs we will now discuss with reference to their cryopreservation. The survival of the brain after circulatory arrest will be discussed below.

10.1.1 Nervous tissue

Nervous tissue can well survive storage at 4°C for a maximum of 8 days (Frodl et al. 1995; Gage et al. 1985; Nikkhah et al. 1995). The first attempts to cryopreserve nervous tissue were published by Luyet and Gonzales in 1953. Since then, various studies have led to other methods. In a summary was given by Paynter (Das et al. 1983; Higgins et al. 2011; Ichikawa et al. 2007; Ma et al. 2010; Paynter 2008; Pichugin 2006a; Quasthoff et al. 2015; Robert et al. 2016).

A number of heterothermic animals outlive freezing temperatures of their bodies and nervous systems (10.4)

10.1.2 The Brain

Hamsters endured freezing of 60 % of their brain water, but the brains were not cooled down to cryogenic temperatures (s.10.3).

Even some free living mammals outlive freezing temperatures of their bodies (19.4)

In 2017, Rodriguez-Martinez et al. proposed cryopreservation of the entire medial ganglionic eminence (MGE). This structure of neural tissue appears only transiently before birth. The structure is located in the wall of the lateral cerebral ventricle. The transmitter GABA is produced here, and this production could be exploited by transplantation of the MGE if there is a deficiency. The explanted MGE tissue forms cells and produces as many cells as fresh tissue even after cryopreservation (Palmero et al. 2016; Purcell et al. 2003; Rahman et al. 2010; Swett et al. 1994).

Neocortical tissue of rat embryos cooled to -70 or -90 °C has been successfully transplanted into the cerebellum of neonatal rats (Das et al 1983; Houlé, Das 1980).

Most studies on cryopreservation of neural tissue examined viability primarily using the state of cell membranes but also based on sprouting and electrical activity of nerve cells with good success after thawing (Pischeda et al. 2018; Rahman et al. 2010). In embryonic tissue from rat brain following cryopreservation, 8-62% of cells present before cooling were alive. Tissue characteristics and, of course, the methods used played a role. Most of the work was done on intact nerve tissue (Luyet, Gonzales 1953; Fang, Zhang 1992; Rodriguez-Martinez et al. 2017). Often, embryonic rat neural tissue is used for scientific studies, and several research groups have studied its cryopreservation. Embryonic cerebral cortical cells are very fragile and survive only 8.2% (Negishi 2002).

Lately brain-region-specific organoids of complex cytoarchitecture 3 mm in diameter, derived from human stem cells, could be cryopreserved and revived at day 21. They were further cultivated up to 150 days. Brain tissue samples of 3mm diameter could also be successfully cryopreserved (Xue et al 2024)

Furthermore, studies have been conducted on cryopreservation of nerve grafts (Bojic et al. 2021; Decherchi et al. 1997; Evans et al. 1998; Fansa 2000; Huang et al. 2018; Jensen et al. 1990; Kohama et al. 2001).

Cryopreserved tissue from the brain area named hippocampus was retransplanted into the brain by Sorensen et al. (1986). Sections of brain tissue are a suited material for studies. Especially, sections containing mature adult brain tissue are interesting e.g. for pharmacology in the context of neuropsychiatry. In 2006, Pichugin et al. proved for the first time that the viability and structure of neural tissue can be well preserved by vitrification of sections. They used a mixture of DMSO, formamide, and ethylene glycol with ice blockers as a cryoprotective solution. Sections from the hippocampus were 0.475 mm thick and thus contained the full organ structure.

Even though it has not yet been possible to vitrify a whole brain, it is thus clear that organ parts of the brain survive vitrification. Preserving a whole brain in this way may be a problem of size.

However, whole cat brains were cooled to -20°C in 15% glycerol (62% of the brain water being present as ice). After 777 days or 7.25 years, they were warmed and showed normal looking electrical brain wave forms (electroencephalograms) in both cases, although activity was decreased after 7.25 years. Hemorrhages and cell loss were also then detected (Suda et al. 1966; -1974). The results have not yet been proven by reproduction.

Rabbit brains perfused with 23% glycerol (3 molar) at room temperature and cooled to dry ice temperature showed excellent preservation on light microscopic observation (Fahy et al. 1984).

Electron-optical findings depend very much on the methods e.g. dehydration and protein precipitation and should be confirmed under other conditions and using other methods (see also Galhuber 2021; Sames 1990; -1994 p.37).

Given the high fat and water content of the brain, it was feared that the very sensitive fine structures, especially the cell connections (synapses), would be difficult to preserve with antifreeze.

However, it has been reported that 80% of cell junctions (synapses) in brain sections retain metabolic properties identical with those in fresh biopsy sections when cooled to -70°C (Hardy et al. 1983). Preservation of viability is as important as preservation of structure.

It also turns out that brain tissue — perhaps even because of its high water and lipid content — and its uptake of large amounts of glucose as well as the possibility of direct viability tests (by electrical brain activity) is particularly suitable for testing cryoprotectants. The results can then also benefit other organs.

A special problem of brain tissue in the living organism is the very difficult replaceability of lost cells, which can lead to severe disorders. Experiments on brain tissue therefore require optimal frost protection. However, this also sets a standard for the highest requirements.

Experiments on brains, especially whole brains, are ethically questionable because it is not known if they are still sentient.

10.2 A market for cryonics: artificially produced tissue and organs

Cryobiology will be needed for so-called regenerative medicine, which addresses the repair and restoration of tissues and organs (Kocsis et al. 2002; Sames 2000a; Toner, Kocsis 2002). This includes repair using stem cells and the development of artificial organs. Recently, tissues have even been produced using 3D printers (Isaacson et al. 2018).

In turn regenerative medicine will be indispensable for the recovery of cryonic patients after rewarming. In addition to the storage of stem cells, cryobiology now also allows, in principle, the storage of artificial tissues and organs over long periods of time.

Cryopreservation of artificial tissue engineering products is a prerequisite for their widespread use. During cryopreservation, the structure and function of the artificial tissues should of course be preserved.

Initial clinical investigations suggest a large demand (Costa et al. 2012). A large market for artificial organs and artificially produced larger pieces of tissue could thus be served. Today, this area still lacks the storage facilities to build up an assortment and easily bring it into commerce.

Artificially created tissues have already been successfully preserved using cryoprotectants. For example. retinal-type nerve tissue grown from human cells (Day et al. 2017; Nakano et al. 2012) and retina models grown from adult stem cells were cryopreserved (Reichman et al. 2017).

So-called organoids have been produced from cultured cells of intestine, liver, heart, brain, kidney and other organs in cultures and can be used for a wide variety of applications e. g., to avoid experiments on animals. Cryopreservation allows their permanent availability, but is not just simple (Pereira et al. 2020).

In artificial skin substitutes from tissue engineering, cryopreservation with 10% DMSO leads to severe loss of cells and delayed healing, i.e. reduced survivability (Harriger 1997). Lamellae composed of skin surface cells, however, can be used to promote wound healing. In 2011, Chen et al. showed that cell lamellae cryopreserved with trehalose could be used for this purpose, as well as fresh membranes (trehalose improved the effect of DMSO in other experiments). Artificial skin substitutes were also cryopreserved in 10% DMSO at a cooling rate of 1°C/min by Wang et al. (2007), and the viability was 75% compared to fresh material.

Artificial esophagi whose cells had been removed were slowly cooled and stored in nitrogen vapor to serve as scaffolds for artificial organ production (Urbani et al. 2017).

Artificial skeletal muscle is needed for reconstructive medicine, robotic movement, disease study and drug testing. The artificial tissue could be frozen and even gain strength during this procedure (Grant et al. 2019).

Artificial bone tissue derived from canine bone marrow and a bone base scaffold partially depleted of minerals was stimulated to form bone and successfully vitrified by Yin et al. (2009). Bone tissue derived from adult stem cells (iPSCs) was also maintained at -80°C by Tam et al. (2020), which unfortunately resulted in cell death. In contrast, the tissue could be maintained alive in a saline solution at 4°C.

In addition to natural grafts (Matsuzaki et al 1996), artificial cartilage tissue has been produced and cryopreserved (Lübke et al 2001). Cartilage can also be produced from cryogenically cooled cells (Gorti et al. 2003; see Bojic et al. 2021 for this section).

10.3 An early success: cooling to freezing temperatures and reanimation

It is indeed possible to cool mammals below 0°C and revive them. The attempts to do so date back a long time and are sparsely known.

Before 1951, Dr. RK Andjus in Belgrad planned a hypothermia experiment with rats. His library had fallen victim to the bombs during the war. So, he could not inform himself and did not know that at that time it was assumed that mammals do not survive temperatures below 15°C alive. He cooled his rats unconcernedly to 2 - 0°C and reanimated them again after 40 - 60 min. Extensively such rat and mouse experiments were continued later at the National Institute of Medical Research in the USA.

Ultimately, the experiments led to the following results.

Golden hamsters (which are the most robust survivors of cooling) were cooled to temperatures below 0°C and rewarmed.

Cooling was performed in a bath of -5°C. The rectal temperature of a hamster that was subcooled to lower temperatures decreased to -0.6°C.

After 40-60 min at -5°C, 15-45% of the water in the body, 53-63% of the cerebrospinal fluid, and 57-90% of the water in the subcutis were in the form of ice.

Hamsters chilled at -5°C for 60 min all recovered when less than 15% of body water was ice. At 15-40% ice, two-thirds recovered, and at 40-50%, one-third recovered. At 55-76% ice in the body, heartbeat and partial respiration could return, but not consciousness. Ice content was recorded by dissection or by calculation from anal temperature.

Some hamsters showed spontaneous heating in the bath of -5°C from 0 to +6°C. They became stiff and crystallized out of the hypothermia, followed by a temperature rise due to the release of the heat of crystallization. The temperature then stabilized to the freezing point of the plasma.

Warming and resuscitation were achieved by diathermy devices or irradiation with a lamp. The internal temperature rose steadily, and the heart began to beat. However, breathing had to be assisted by instilling air until it resumed on its own. No frostbite appeared. However, animals kept at -5°C for longer than one hour did not recover.

Twenty-five percent of the animals in a -5°C bath showed a drop in temperature below freezing. In some experiments, the temperature was lowered to -8°C, and the animals were able to recover completely.

Hamsters that crystallized at -2.8°C recovered completely. If crystallization occurred at -3 to -6°C, they rarely recovered, despite a transiently detectable heartbeat.

J.E. Lovelock had developed the diathermy apparatus used for heating, which produced an alternating magnetic field. The heating rate was 16°C per minute.

The experiments were repeated on rabbits and small primates (galagos a semi-monkey species), but all these animals died a few hours after rewarming. A number of rabbits and galagos recovered after the temperature dropped to the freezing point.

It has been shown that hamsters, who survive do not suffer any loss in behavioral range after rewarming. The successes were explained by the normal behavior of the tissue during slow cooling by the freeze-thaw method (Goldzweig, Smith 1956; Lovelock, Smith 1956; -1958; Smith 1961; Smith et al. 1954).

Cryostasis of whole animals naturally preserves different organs. This complements the results already discussed on isolated organs.

The following formula was found for vitrification, assuming 68% glycerol, because it was survived by hamsters.

$$C = 68 - 0.68\,(P)$$

C is the tolerable glycerol concentration, and P is the percentage of the liquid content of a tissue that can be converted to ice without damage by crystals (Fahy et al. 1987).

If brain tissue can tolerate a 60% conversion of its water to ice, the formula is C = 68 - 0.68 (60) = 27.2.

Thus, 27.2% glycerol should be sufficient for deep-cooling the brain at any temperature without damage from ice crystals.

10.4 Nature a teacher for cryopreservation — is it better than our laboratories?

Most people are not necessarily aware of it anymore, but remember: we were told in school (e.g., in the 1950s) that in polar regions and countries with cold winters, animals and plants can survive in amazing ways by dropping their temperature to freezing degrees (Ross 1990; Fuller et al. 2004).

Since cryopreservation already occurs in nature, it is clear that there are conditions at low body temperatures below 0°C that are compatible for all organs of small animals. However, no temperatures occur in the nature of the earth that are as low as cryogenic temperatures such as our glass transition temperature which are optimal for long times of storage. In nature, low temperatures do not last even for a whole year even in most of the inhabited zones.

It seems like a miracle that, on the other hand, some much less developed tiny animals in nature could be revived after thousands of years in ice. This fact shows that the damage caused by storage at temperatures above the glass transition temperature must be reexamined.

Rotifers, which can be counted as worms, have been found in layers of Siberian permafrost dated to 24 000 years old. Temperatures here are likely to be well above -130°C around -22 to -36°C (Obu et al. 2020). Nematodes were found in 30000-year-old strata. Both species have been revived. Simply built small creatures can thus survive tens of thousands of years at relatively high freezing temperatures (Shmakova 2021).

In contrast, warm-blooded animals are very sensitive to cold. For mammals, 15°C is considered the lethal limit of hypothermia.

Their brains function only at virtually full body heat. Surprisingly, Arctic ground squirrels (Barnes, Buck 2000; Popovic, Popovic 1963) and other mammals can cool below freezing during hibernation. Ruf and Geiser (2015) found 8 mammalian species that reached body temperatures below 0°C during hibernation, including 3 in which they fell below -2°C. Most warm-blooded mammals however, cope with cold only by regulating body temperature. In comparison cold-blooded animals rely on other means.

The antifreeze substances of animals that survive hibernation or cooling below 0°C belong to the classes of single sugars and polymeric sugars as well as proteins and lipids. So, there are very different ones. Binding sites to ice have been studied for proteins (Graham et al. 2008). A wide range of antifreeze proteins exists in fish, insects, worms, plants, and bacteria (Tas et al. 2021). Various species of fish, amphibians, and reptiles survive the transition of their body water to the solid phase with the help of antifreeze agents and the production of so-called antioxidants, substances that render free radicals and toxic oxygen products harmless. Regulations in the field of molecules and their metabolism interact with antifreeze agents in the survival of freezing temperatures (e. g. Storey, Storey 2001).

Cold-blooded (heterothermic) animals have a different heat balance compared to warm-blooded animals and the corresponding organization. Their brains, for example, can still function at temperatures as low as 0°C. The natural antifreeze processes in heterothermic vertebrates such as amphibians and reptiles have been described in detail (Costanzo et al. 1995; Lowe et al. 1971; Storey 1990; -1999; Storey, Storey 1993).

Reptiles can endure at least brief periods of supercooling, as demonstrated by the wood lizard. It occurs from western Europe to the Arctic ring (Costanzo et al. 1995a).

Frogs can spend days or weeks while 65% (not more) of their body water is ice. Some amphibians use glycerol (which is indeed produced in all cells, especially in the liver) as frost protection. The liver releases it into the body.

Nordic frogs, for example, use glucose as a cryoprotectant, which is formed in the liver when temperatures drop (no lower than -6°C) and is channeled into the cells in large quantities by a special form of the hormone insulin (Conlon M et al. 1998). The heart and brain do not freeze, but other organs do (the water in a frog is 2/3 ice in winter). Freezing in the wood frog (ice frog) Rana sylvatica is gradual, starting with the legs and increasing as it gets colder into the abdominal cavity, around the abdominal organs (Costanzo et al. 2015; Lee et al. 1992; Storey 1997).

Various insects and fish species also use glucose or glycerol, while others use short protein chain antifreeze peptides (called AFPs). The latter cause a non-colligative antifreeze effect, which is much greater than one would expect from the concentration, because they not only lower the freezing point, but also hinder ice crystal formation directly even in the cells. They also stabilize the cell membrane. 5 types of such AFPs have been found in fish and two in insects (table in Bruinsma, Uygun 2017).

A fish antifreeze protein was tested on botanical material under vitrification conditions (Wang et al. 2001).

There have been many studies on the various naturally occurring antifreeze proteins (Davies et al. 2002; Graham et al. 1997; Graham et al. 2008; Zhang L et al. 2022)

We often consume such proteins with food, but there are no known effects or side effects (Crevel et al. 2002). We do not know whether a cryonic patient should eat many cold water fish before death, for example. However, proteins are usually simply digested in the body.

Invertebrates (Storey and Storey 1992, 1996) have evolved diverse methods for enduring frost even alongside the formation of different antifreeze agents.

Here a fact turns out, which leads the scientific knowledge a piece further namely that nevertheless whole animals (even if small) can be cooled down to temperatures, which do not occur at all in the nature and namely down to the temperature of liquid helium or even quite near the absolute zero point. These are in particular insects, worms and tardigrades.

Proteins can lower the freezing point so that its temperature shows a difference from the temperature of the melting point (thermal hysteresis), delaying the formation of ice during cooling (antifreeze proteins are thermal hysteresis proteins).

Some insects have adapted well to freezing temperatures. There are Central European butterflies such as the brimstone butterfly that can survive unprotected winters, with their bodies cooling to subzero temperatures (Turnock, Fields 2005).

In 1997, a protein was obtained from the common mealworm beetle Tenebrio Molitor that is 100 times more effective than fish antifreeze proteins. One milligram of this protein in one milliliter of solution can lower the freezing point by 5.5°C (Graham et al. 1997). Another frost-resistant beetle is the arctic beetle Upis ceramboides (Walters et al. 2009). The larva of an American/Canadian bark beetle nicely named Cucujus clavipes puniceus undergoes a transformation to a glassy state at a temperature of -58°C and can avoid forming ice crystals down to at least -150°C. In the adult beetle (the imago) itself, this temperature is -100°C (Sformo et al. 2010). The processes involved are currently being researched in this species to elicit its tricks. This much is already certain: it has known a few tricks of the cryonicists a long time before and perhaps more, especially the avoidance of ice crystals that act as nuclei through antifreeze proteins and the formation of polymer sugars with lipids as side chains (lipopolysaccharides). It also cleans its gut to eliminate (exogenous) crystallization nuclei and enters a resting stage (Bennett et al. 2005; Carrasco et al. 2011; - 2012; Duman et al. 1984). Another Arctic beetle the ground beetle Perostichus brevicornis can tolerate temperatures below-35°C; in the laboratory it was cooled to -87°C for 5 hours without apparent damage (Miller 1969). We also mention here the fire beetle Dendroides Canadensis, whose larvae have been shown to survive -3°C (Nickell et al. 2013). An antifreeze protein of the desert beetle Anatolica polita can protect eggs and embryos of the frog Xenopus laevis from cold (Jevtić et al. 2022).

Enduring Arctic insects also use proteins that inhibit c-axis growth. They can lower the freezing point by 4-5°C (Pertaya et al. 2008).

Thus, the enduring of low temperatures is not an isolated case in insects, but a widespread principle of survival in winter.

It is interesting to investigate what the antifreeze proteins do with the water. This can be observed with a tetrahertz spectrometer. In the vicinity of the antifreeze proteins, water molecules, which are normally in constant motion and bump each other, slow down to bond very briefly and then break away again. The bonds take a little longer with antifreeze protein. The slower this occurs, the lower the temperature at which the water freezes (Meister et al. 2013; -2014).

In the late 17th and early 18th centuries, Van Leeuwenhoek found, among other things, so-called "animaliculi", tiny animals such as nematodes, rotifers or tardigrades, in deposits on house roofs in the "moss lawn". He himself called them "Diertgens." The animals can survive the winter there (see Lauritz 1996).

Worms and worm-like animals probably master ice survival most perfectly.

The 2.2- to 7- mm long-leech Ozobranchus jantseanus can survive temperatures as low as -196°C (liquid nitrogen temperature) without any special treatment (Suzuki et al. 2014).

Nematodes are even smaller (Roth, Nystul 2005; Wharton et al. 2005). The tiny nematode Caenorhabditis elegans, in whose organs age-stable proteoglycans can be detected (Schimpf 1997; Schimpf et al. 1999), survives cooling to the temperature of liquid nitrogen (Bird, Bird 1991; Saul 2018). Another nematode, vinegar fluke (Turbatrix aceti) was first described in 1656. Luyet and Gehenio reported in 1950 that it survives cooling in liquid nitrogen.

Rotifers (rotifers; King et al. 1983; Toledo, Kurokura 1990) survive the same temperature.

We have already mentioned that such animals also survive millennia on ice. Even more resistant are the equally small tardigrades (Tardigrada) especially their permanent forms. These animals can be cooled to - 196°C (Ramlov, West 1992). However, they could be cooled even deeper as deep as possible in the laboratory to near absolute zero. Therefore, they can survive without energy as a pure structure. They survived the conditions of near-Earth space to which they were openly exposed (Jönsson et al. 2008). For cryonics, they prove that animals with all major organs can survive temperatures of liquid helium. With a lot of imagination, they could be taken to be aliens from outer space, because they have properties that they cannot fully exploit on Earth. These would then be the only space migrants we have had contact with so far (Kletetschka et al. 2015; Schill 2018). They can also tolerate very high temperatures, UV radiation, vacuum, and desiccation. Nematodes and tardigrades survive desiccation with the help of trehalose (Crowe et al. 2001).

Plants can vitrify at temperatures of -30 to -40° C (Hirsh 1987). Such plants could be tested as a source for obtaining natural antifreeze. Sucrose is the sugar most commonly found in frost-resistant plants. These can actually increase their sucrose content 10-fold at low temperature.

Fungi also engage in frost protection (Kawahara et al. 2016), and bacteria have already evolved ways to survive frost (Chattopadhyay 2005; Cid et al. 2016). Thus, these capabilities are likely ancient.

10.4.1 Hibernation

Hibernation like cryopreservation is a matter of protection against frost and dangers due to crystal formation. However, other possibilities are used to some extent than in laboratories and the conditions are different. The time is limited to one winter and the temperatures remain relatively high. The processes during hibernation in mammals have been described in detail (Storey 2005). In hibernating Arctic squirrels, the number of white blood cells drops a hundredfold. This prevents them from sticking to the wall of blood vessels and occluding them (Drew al. 2001) (This sticking of blood cells is one of the reasons that a human brain cannot be resuscitated after 5-9 minutes). The squirrel heart rate can be slowed hundreds of times and the metabolic rate can be reduced to 5% of normal. For small hibernating mammals, the temperature can drop to near 0°C (in contrast, in non-hibernating mammals, the heart stops at 10-20°C). Passive exchange of potassium and sodium across the cell membrane is reduced. Calcium exclusion from the cell is increased. Energy is derived from catabolic processes without oxygen (Carey et al. 2003). The brains of hibernating ground squirrels appear to be particularly stable compared to other rodents especially when removed during hibernation. The Brains of animals that were perfused were able to maintain their capacity to produce electrical activity for three days (Pakhotin, Pakhotina 1994; Pakhotin et al. 1990; -1993).

There exists widespread agreement among researchers that the human species is incapable of hibernation and that this is genetically determined by our tropical origins. However, insect eaters close to the primate root such as hedgehogs and even tropical Tanreks are hibernators. In addition, certain tropical primates are capable of hibernation. This has been demonstrated in Malagasy dwarf lemurs e.g. the fat-tailed lemur Cheirogaleus medius (Dausmann et al. 2004, -2009). Since humans belong to the primates, hibernators could even be at our roots, at least if one goes back to those insect eaters from which primates are descended.

We also need to remember here that our tissues are full of natural antifreeze. There are a number of polymeric sugars and other sugars in our bodies, in addition to the proteoglycans already mentioned, that can act as cryoprotectants. Animals, for example, use simple glucose to survive winter freezing. It is abundant in the human brain.

Unfortunately, getting the temperature below 0°C during hibernation without damage remains a problem.

The same applies to animals that cool down to minus temperatures in winter. Unfortunately, they can hardly be cooled lower without damage (with the exception of the invertebrates discussed above, such as some insects, tardigrades or worms).

Mammals (and probably also humans), which can be cooled down to 10°C by the so-called suspended animation (see above), also have not been cooled further from this state so far.

Scientific tests must show whether a method can be developed to achieve this anyway. In particular, cryoprotectant would have to be used together with deoxygenation and cooling.

Overall, this chapter highlights encouraging opportunities for cold survival that may be developed in a short period of time.

Literature

Barnes BM, Buck C (2000) Hibernation in the Extreme: Burrow and body temperatures, metabolism, and limits to torpor bout length in arctic ground squirrels. In Heldmaier G, Klingenspor M (eds) Life in the cold. Springer, Heidelberg, pp 65-72

Bennett VA et al (2005) Comparative overwintering physiology of Alaska and Indiana populations of the beetle Cucujus clavipes (Fabricius): roles of antifreeze proteins, polyols, dehydration and diapause. J Exp Biol (JEB) 208:4467-4477

Bird J, Bird AC (1991) The structure of Nematodes. Academic Press, Boston

Bojic S et al (2021) Winter is coming: the future of cryopreservation. BMC Biol 19.56

Bruinsma BG, Uygun K (2017) Subzero organ preservation; the dawn of a new ice age? Curr Opin Organ Transplant 22:281-6

Carey HV et al (2003) Mammalian hibernation: cellular and molecular responses to depressed metabolism and low temperature. Physiol Rev 83:1153-1181

Carrasco MA et al (2011) Elucidating the biochemical overwintering adaptations of larval Cucujus clavipes puniceus, a nonmodel organism, via high throughput proteomics. J Proteome Res 10:4634-4646

Carrasco MA et al (2012) Investigating the deep supercooling ability of an Alaskan beetle, Cucujus clavipes puniceus, via high throughput proteomics. J Proteomics 75:1220-1234

Chattopadhyay MK (2005) Mechanism of bacterial adaptation to low temperature. J Biosci 31:157-165

Chen F et al (2011) Cryopreservation of tissue-engineered epithelial sheets in trehalose. Biomaterials 32:8426-8435

Cid FP et al (2016) Properties and biotechnological applications of ice-binding proteins in bacteria. FEMS Microbiol Lett 363:1-12

Conlon M et al (1998) Freeze tolerance in the wood frog Rana sylvatica is associated with unusual structural features in insulin but not in glucagon. J Mol Endocrinol 2:153-15

Costa PF et al (2012) Cryopreservation of cell/scaffold tissue-engineered constructs. Tissue Eng Part C 18:852-858

Costanzo JP et al (1995) Survival mechanisms of vertebrate ectotherms at subfreezing temperatures: applications in cryomedicine. FASEB J 9:351-358

Costanzo JP et al (1995a) Supercooling, ice inoculation and freeze tolerance in the European common lizard, Lacerta vivipara. J Comp Physiol B 165:238-244

Costanzo JP et al (2015) Cryoprotectants and extreme freeze tolerance in a subarctic population of the wood frog. PLoS One. 10:e0117234

Crevel RW et al (2002) Antifreeze proteins: occurrence and human exposure. Food Chem Toxicol 4:899-903

Crowe JH et al (2001) The trehalose myth revisited: introduction to a symposium on stabilization of cells in the dry state. Cryobiology 43:89-105

Das GD et al (1983) Freezing of neural tissues and their transplantation in the brain of rats: technical details and histological observations. J Neurosci Methods 8:1-15

Dausmann KH et al (2004) Physiology: hibernation in a tropical primate. Nature 429:825-826

Dausmann KH et al (2009) Energetics of tropical hibernation. J Comp Physiol B 179:345-357

Davies PL et al (2002) Structure and function of antifreeze proteins. Philos Trans R Soc Lond B Biol Sci 357:927-935

Day AGE et al (2017) The effect of hypothermic and cryogenic preservation on engineered neural tissue. Tissue Eng Part C Methods 23:575-582

Decherchi P et al (1997) CNS axonal regeneration with peripheral nerve grafts cryopreserved by vitrification: cytological and functional aspects. Cryobiology 34:214-239

Drew KL et al (2001) Neuroprotective adaptations in hibernation: therapeutic implications for ischemia-reperfusion, traumatic brain injury and neurodegenerative diseases. Free Radic Biol Med 31:563-573

Duman JG et al (1984) Change in overwintering mechanism of the cucujid beetle, Cucujus clavipes. J Insect Physiol 30:235-223

Evans PJ et al (1998) Cold preserved nerve allografts: changes in basement membrane, viability, immunogenicity, and regeneration. Muscle Nerve 21:1507-1522

Fahy GM et al (1984) Vitrification as an approach to cryopreservation. Cryobiology 21:407-426

Fahy GM et al (1987) Some emerging principles underlying the physical properties, biological actions, and utility of vitrification solutions. Cryobiology 24:196-213

Fang J, Zhang ZX (1992) Cryopreservation of embryonic cerebral tissue of rat. Cryobiology 29:267-273

Fansa H (2000) Cryopreservation of peripheral nerve grafts. Muscle Nerve 23:1227-1233

Frodl EM et al (1995) Effects of hibernation or cryopreservation on the survival and integration of striatal grafts placed in the ibotenate-lesioned rat caudate-putamen. Cell Transplant 4:571-577

Fuller BJ et al (2004) Life in the frozen state. CRC Press, Boca Raton

Gage FH et al (1985) Rat fetal brain tissue grafts survive and innervate host brain following five day pregraft tissue storage. Neurosci Lett 60:133-137

Galhuber M et al (2021) Simple method of thawing cryo-stored samples preserves ultrastructural features in electron microscopy. Histochem Cell Biol 155:593-603

Goldzweig SA, Smith AU (1956) A simple method for reanimating ice-cold rats and mice. J Physiol 132:406-413

Gorti GK et al (2003) Cartilage tissue engineering using cryogenic chondrocytes. Arch Otolaryngol Head Neck Sur 129:889-889

Graham LA et al (1997) Hyperactive anrtifreeze protein from beetles. Nature 388:727-728

Graham LA et al (2008) Hyperactive antifreeze protein from fish contains multiple ice-binding sites. Biochemistry 47:2051-2063

Grant L et al (2019) Long-term cryopreservation and revival of tissue engineered skeletal muscle. Tissue Eng Part A 25:1023 1036

Hardy JA et al (1983) Metabolically active synaptosomes can be prepared from frozen rat and human brain. J Neurochem 40:608-614

Harriger MD et al (1997) Reduced engraftment and wound closure of cryopreserved cultured skin substitutes grafted to athymic mice. Cryobiology 35:132-142

Higgins AZ et al (2011) Effects of freezing profile parameters on the survival of cryopreserved rat embryonic neural cells. J Neurosci Methods 201:9-16

Hirsh AlG (1987) Vitrification in plants as a natural form of cryoprotection. Cryobiology 24:214-228

Houlé JD, Das GD (1980) Freezing and transplantation of brain tissue in rats. Experientia 36:1114-1115

Huang YY et al (2018) Various changes in cryopreserved acellular nerve allografts at -80 degrees C. Neural Regen Res 13:1643-1649

Ichikawa J et al (2007) Cryopreservation of granule cells from the postnatal rat hippocampus. J Pharm Sci 104:387-391

Isaacson A et al (2018) 3D bioprinting of a corneal stroma equivalent. Exp Eye Res 173:188-193

Jevtić P et al (2022) An insect antifreeze protein from Anatolica polita enhances the cryoprotection of Xenopus laevis eggs and embryos. J Exp BIOL 225:jeb243662. doi: 10.1242/jeb.243662

Jensen S et al (1990) Cryopreservation of rat peripheral nerve segments later used for transplantation. Neuroreport 1:243-246

Kawahara H et al (2016) Antifreeze activity of xylomannan from the mycelium and fruit body of Flammulina velutipes. Biocontrol Sci 21:153-159

King CE et al (1983) Cryopreservation of monogonont rotifers. In Pejler B et al (eds) Biology of Rotifers. Developments in Hydrobiology, vol. 14, Springer, Dordrecht, https://doi.org/10.1007/978-94-009-7287-2_12

Kletetschka G, Hruba J (2015) Dissolved gases and ice fracturing during the freezing of a multicellular organism: lessons from Tardigrades. Biores 4:209-217

Kocsis JD et al (2002) Storage and translational issues in reparative medicine: breakout session summary. Ann N Y Acad Sci 961:276-278

Kohama I et al (2001) Transplantation of cryopreserved adult human Schwann cells enhances axonal conduction in demyelinated spinal cord. J Neurosci 21:944-950

Lauritz S (1996) Anhydrobiosis and cold tolerance in Tardigrades. Eur J Entomol 93:349-375

Lee RE Jr, Costanzo JP et al (1992) Dynamics of body water during freezing and thawing in a freeze-tolerant frog (Rana sylvatica). J Therm Biol 17: 263-266

Lovelock JE, Smith AU (1956) Studies on golden hamsters during cooling to and rewarming from body temperatures below 0 degrees C. III. biophysical aspects and general discussion. Proc Royal Soc B 145:427-442

Lowe CH et al (1971) Supercooling in reptiles and other vertebrates. Comp Biochem Physiol A Comp Physiol 3:125-135

Lübke C et al (2001) Cryopreservation of artificial cartilage: viability and functional examination after thawing. Cells Tissues Organs 169:368-376

Luyet BJ, Gehenio PM (1950) Survival of vinegar eels after congelation in liquid nitrogen. Anat Rec 108:544

Luyet B, Gonzales F (1953) Growth of nerve tissue after freezing in liquid nitrogen. Biodynamica 7:171-174

Ma XH et al (2010) Slow-freezing cryopreservation of neural stem cell spheres with different diameters. Cryobiology 60:184-191

Matsuzaki H et al (1996) Allografting Intervertebral Discs in Dogs A Possible Clinical Application. Spine 2:178-183

Meister K et al (2013) Long-range protein-water dynamics in hyperactive insect antifreeze proteins. Proc Natl Acad Sci 110:1617-1622

Meister K et al (2014) The role of sulfates on antifreeze protein activity. J Phys Chem B 118:7920-7924

Miller LK (1969) Freezing tolerance in an adult insect. Science 166:105-106

Nakano T et al (2012) Self-formation of optic cups and storable stratified neural retina from human ESCs. Cell Stem Cell 10:771-785

Negishi T (2002) Cryopreservation of brain tissue for primary culture. Exp Anim 51:383-390

Nickell PK et al (2013) Antifreeze proteins in the primary urine of larvae of the beetle Dendroides canadensis. J Exp Biol 216:1695-1703

Nikkhah G et al (1995) Preservation of fetal cells by cool storage: in vitro viability and TH-positive neuron survival after microtransplantation to the striatum. Brain Res 687:22-34

Obu et al (2020) Pan-Antarctic map of near-surface permafrost temperatures at 1 km2 scale. The Cryosphere 14:497-519

Pakhotin PI, Pakhotina ID (1994) Preparation of isolated perfused ground squirrel brain. Brain Res Bull 33:719-721

Pakhotin PI et al (1990) Functional stability of the brain slices of ground squirrels, Citellus undulatus, kept in conditions of prolonged deep periodic hypothermia: electrophysiological criteria. Neuroscience 38:591-598

Pakhotin PI et al (1993) The study of brain slices from hibernating mammals in vitro and some approaches to the analysis of hibernation problems in vivo. Prog Neurobiol 40:123-161

Palmero E et al (2016) Brain tissue banking for stem cells for our future. Sci Rep 6 Art No: 39394

Paynter SJ (2008) Principles and practical issues for cryopreservation of nerve cells. Brain Res Bull 75:1-14

Pereira EC et al (2020) Principles of cryopreservation and applicabilities in intestinal organoids. Japanese Journal of Gastroenterology and Hepatology, DOI:10.47829/JJGH.2020.41103; JJGH-v4-1391

Pertaya N et al (1963) Survival of newborn ground squirrels after supercooling or freezing. Am J Physiol 204:949-952

Pertaya N et al (2008) Direct visualization of spruce budworm antifreeze protein interacting with ice crystals: basal plane affinity confers hyperactivity. Biophys J. 95:333-341

Pichugin Y et al (2006) Cryopreservation of rat hippocampal slices by vitrification. Cryobiology 52:228-240

Pichugin Y (2006a) Problems of long-term cold storage of patients' brains for shipping to CI. The Immotalist 38:14-20

Pischedda F et al (2018) Cryopreservation of primary mouse neurons: the benefit of neurostore cryoprotective medium. Front Cell Neurosci 12:81

Popovic P, Popovic V (1963) Survival of newborn ground squirrels after supercooling or freezing. Am J Physiol 204:949-952

Purcell WM et al (2003) Cryopreservation of organotypic brain spheroid cultures. Altern Lab Anim (ATLA) 31:563-573

Quasthoff K et al (2015) Freshly frozen E18 rat cortical cells can generate functional neural networks after standard cryopreservation and thawing procedures. Cytotechnology 67:419-426

Rahman AS et al (2010) Cryopreservation of cortical tissue blocks for the generation of highly enriched neuronal cultures. J Vis Exp (JoVE) 45:238

Ramlov H, West P (1992) Survival of the cryptobiotic eutardigrade Adorybiotis coronifer during cooling to -196°C effect of cooling rate, trehalose level, and shortterm acclimation. Cryobiology 29:125-130

Reichman S et al (2017) Generation of storable retinal organoids and retinal pigmented epithelium from adherent human iPS cells in xeno free and feeder-free conditions. Stem Cells 35:1176-1188

Robert MC et al (2016) Cryopreservation by slow cooling of rat neuronal cells. Cryobiology 72:191-197

Rodriguez-Martinez D et al (2017) Cryopreservation of GABAergic neuronal precursors for cell-based therapy. PLoS One. 2017; 12: e0170776, doi: 10.1371/journal.pone.0170776

Ross PE (1990) Cold storage. Winter-proof critters suggest ways to store human. Sci-Am 26:20-21

Roth MB, Nystul TG (2005) Survival in cold sleep. Spectrum Wiss, Sept 2005:42-48

Ruf T, Geiser F (2015) Daily torpor and hibernation in birds and mammals. Biol Rev 90:891-962

Sames K (1990) Age related changes of morphological parameters in hyaline cartilage. In Robert L, Hofecker G (eds) Vienna Aging Series, vol. 2, Facultas, Vienna, pp 177-84

Sames K (1994) The role of proteoglycans and glycosaminoglycans in aging. In Hahn HP (ed) Interdisciplinary Topics in Gerontology, vol. 28. Karger, Basel

Sames K (ed) (2000a) Medical Regeneration and Tissue Engineering. Ecomed, Landsberg

Saul N (2018) Anti-aging and pro-longevity: what can we learn from a small worm? A methodical overview. In Sames KH (ed) Applied Human Cryobiology, vol. 2, ibidem, Stuttgart

Schill RO (ed) (2018) Water Bears: The Biology of Tardigrades. Zoological Monographs, Springer Nature Switzerland AG

Schimpf J (1997) Altern und alternsabhängige Veränderungen polyanionischer Strukturen bei der Nematode Caenorhabditis elegans. Dissertation, University of Erlangen

Schimpf J et al (1999) Proteoglycan distribution pattern during aging in the nematode Caenorhabditis elegans: an ultrastructural histochemical study. Histochem J 31:285-292

Sformo T et al (2021) Deep supercooling, vitrification and limited survival to -100°C in the Alaskan beetle Cucujus Clavipes Puniceus (Coleoptera: Cucujidae) Larvae. J Exper Biol 213:502-509

Shmakova L et al (2021) A living rotifer from 24,000-year old arctic permafrost. Curr Biol 31:R712-R713

Smith AU (1956) Resuscitation of hypothermic, supercooled and frozen mammals. Proc R Soc Med 49:357-358

Smith AU (1958) Resuscitation of frozen mammals. New Scientist 4:1154

Smith AU (1961) Revival of mammals from body temperatures below zero. In Smith AU (ed) Biological Effects of Freezing and Supercooling, Edward Arnold, London, pp 304-368

Smith AU et al (1954) Resuscitation of hamsters after supercooling or partial crystallization at body temperatures below 0°C. Nature 173:1136-1137

Sorensen T et al (1986) Intracephalic transplants of freeze-stored rat hippocampal tissue. J Comp Neurol 252:468-482

Storey KB (1990) Life In adaptive strategies for natural freeze tolerance in amphibians and reptiles. Am J Physiol 258:R559-R568

Storey K (1997) Organic solutes in freezing tolerance in a frozen state. Com. Biochem Physiol 117A:319-326

Storey KB (1999) Living in the cold: freeze-induced gene responses in freeze-tolerant vertebrates. Clin Exp Pharmacol Physiol 26:57-63

Storey KB (2005) Hibernating mammals. In Sames KH et al (eds) Extending the lifespan biotechnical, gerontological, and social problems. Lit, Münster, pp 219-228

Storey B, Storey JM (1992) Biochemical adaptations for winter survival in insects. In Steponkus PL (ed) Advances in Low Temperature Biology. Vol. 3 Elsevier, Amsterdam pp 101-140

Storey B, Storey JM (1993) Cellular adaptations for freezing survival by amphibians and reptiles. In Steponkus PL (ed) Advances in Low Temperature Biology, vol. 2, Elsevier, Amsterdam, pp 101-129

Storey B, Storey JM (1996) Natural freezing survival in animals. Ann Rev Ecol Syst 27:365-386

Storey KB, Storey JM (eds) (2001) Cell and molecular responses to stress, vol. 2, Elsevier Press, Amsterdam, pp 263-287

Suda I et al (1966) Viability of long term frozen cat brain in vitro. Nature 212:268-270

Suda I et al (1974) Bioelectric discharges of isolated cat brain after revival from years of frozen storage. Brain Research 70:527-531

Suzuki D et al (2014) A leech capable of surviving exposure to extremely low temperatures. PLoS ONE 9:348-349

Swett JW et al (1994) Quantitative estimation of cryopreservation viability in rat fetal hippocampal cells. Exp Neurol 129:330-334

Tam E et al (2020) Hypothermic and cryogenic preservation of tissue-engineered human bone. Ann NY Acad Sci 1460:77-87

Tas RP et al (2021) From the freezer to the clinic. EMBO Rep 22 e52162

Toledo JD, Kurokura H (1990) Cryopreservation of the euryhaline rotifer Brachionus plicatilis embryos. Aquaculture 91:385-394

Toner M, Kocsis J (2002) Storage and transitional issues in reparative medicine. Ann NY Acad Sci 981:258-262

Turnock WJ, Fields PG (2005) Winter climates and coldhardiness in terrestrial insects. Eur J Etomol 102:561-576

Urbani L et al (2017) Long-term cryopreservation of decellularised oesophagi for tissue engineering clinical application. PloS one 12:e0179341

Walters KR et al (2009) A nonprotein thermal hysteresis-producing xylomannan antifreeze in the freezetolerant Alaskan beetle Upis ceramboides. Proc Natl Acad Sci 106:20210-20215

Wang JH et al (2001) The dual effect of antifreeze protein on cryopreservation of rice (Oryza sativa l.) embryogenic suspension cells. Cryo-Letters 22:175-182

Wang X et al (2007) Cryopreservation of tissue-engineered dermal replacement in Me2SO: Toxicity study and effects of concentration and cooling rates on cell viability. Cryobiology 55:60-65

Wharton DA et al (2005) Ice-active proteins from the antarctic nematode panagrolaimus davidi. Cryobiology 51:198-207

Xue W (2024) Effective cryopreservation of human brain tissue and neural organoids. Cell Rep Methods 4:article 100777

Yin H et al (2009) Vitreous cryopreservation of tissue engineered bone composed of bone marrow mesenchymal stem cells and partially demineralized bone matrix. Cryobiology 59:180-187

Zhang L et al (2022) Bioinspired ice-binding materials for tissue and organ cryopreservation. J Am Chem Soc 144:5685-5701

11 The body during oxygen deficiency

Summary

When the circulation stops, there is a general lack of oxygen. The energy runs out. The cells try to survive and adjust their metabolism. Without energy, this does not succeed.

Cryonics can only be performed after a more or less prolonged lack of oxygen (ischemia). The processes during this period therefore create important conditions for the success of cryopreservation of the deceased. The continuation of the changes is responsible for the fact that resuscitation becomes more and more difficult as time passes. The changes continue even in cryonics patients whose organs are no longer functioning.

11.1 Ions and the electropotential of the normal, living cell

It is about understanding the effects of oxygen. Oxygen is the most important energy supplier for life processes. Especially electrical properties of cells cause high energy consumption.

The electrical charge at the cell membrane comes mainly from the distribution of ions and is especially important for nerve cells. Cells and their surroundings contain many negatively charged atoms or molecules (anions) mostly bound to proteins and organic phosphorus compounds as well as proteoglycans. Many large molecules are unable to penetrate cells. On the other hand, even the simple positive calcium ions cannot enter the cells from the outside—although they are small—and also remain in the environment of the cells.

There are channels in the cell membrane for potassium, sodium and chloride. The influx of potassium into the cell is thereby favored, so that there is more potassium than sodium within the cell. The sodium ion binds more water than the potassium ion. As a result, it goes through the channels more slowly and there is more sodium than potassium outside the cell. So, you have more sodium on the outside and more potassium in the inside.

There are also many negatively charged ions in the cell. The positive potassium and sodium ions are basically drawn into the cell by these negatively charged ions. Therefore, potassium does not leave the cell even though there is an excess of potassium inside compared to outside. However, potassium also does not completely balance the negative charges inside the cell. The inside of the cell remains negative. The outside of the cell, on the other hand, is positively charged.

As a result, there is an electrical voltage across the membrane of the cell, which is called membrane potential. One can calculate the voltage generated by a particular ion with an equation (according to Nernst). However, the membrane voltage is caused by the distribution of all ions. Thereby potassium essentially determines the voltage. It can leave the cell through potassium channels, thereby the inside of the cell becomes more negative and the outside more positive. The voltage can collapse in response to a stimulus. This occurs through a sodium channel, which is opened when the voltage changes. Temporarily, more sodium then flows into the cell. This process is called depolarization. Depolarization can occur quite suddenly and is used as a signal by nerve cells.

Such a signal is then transmitted from excited nerve cells to other nerve cells until it reaches the target organ, which goes into action in response to the excitation, for example, a muscle, which contracts when it receives a nerve signal. Thus, by small changes in the concentration of solute molecules, the cell produces enormous and vital effects.

The membrane voltage is maintained by the so-called sodium/potassium pump. Thus, after voltage equalization (depolarization), the original distribution of ions and thus the original voltage is restored. So, to speak, the battery is recharged to generate the next signal. If this "pump" does not work, the membrane voltage may slowly equalize (over hours) because the ions migrate. Then there is no more membrane voltage. The cell is no longer excitable. The pumping of molecules is thus one of the most important life processes, not only in nerve cells.

Most of the energy is actually used by nerve cells to maintain the distribution of ions inside and outside the cell membrane.

The sodium/potassium pump keeps potassium low outside the cell and higher inside the cell.

> Background information
> The pump is driven by the energy generated by the respiratory organelles through a stepwise combination of oxygen with hydrogen. This reaction is controlled by enzymes. The energy generated is stored mainly in the form of phosphate compounds with adenosine. Another source of energy is glucose. One molecule of glucose can form 2 molecules of adenosine triphosphate as energy storage without the direct use of oxygen. However, lactate (salt of lactic acid) is formed as a by-product. This can lead to acidification of the blood. It is generally known that glucose is a readily usable source of energy, and blood acidification by lactate is known from peak athletic performance or heart failure.

Another pump helps maintain ten times the calcium ions (Ca2+) outside the cell, compared to inside the cell.

> Background information
> The distribution of ions is also regulated by direct so-called ion exchangers in the cell membrane.
> The voltage balance travels as a signal through the nerve cell to its processes (extensions). When the signal arrives at the end of a process, chemical transmission is commonly initiated. Certain substances (transmitters) thereby pass through the gap between the extensions of two nerve cells as transmitters. They cause depolarization in the neighboring cell, whereby the excitation is transmitted to the neighboring cell. Here, the influx of calcium into the cell plays a role. The transmitter substance glutamate (one of many), for example, serves this purpose. Glutamate is an important transmitter that is released at cell borders during excitation. There are various receivers (receptors) for glutamate at the membrane of the neighboring cell. When glutamate is bound to them, these receptors allow calcium ions to enter the cell. These are namely so-called NMDA and AMPA receptors. In addition, there are "voltage-gated" calcium channels in the membrane, in which an L-type and a T-type are distinguished. Furthermore, there is an exchanger for sodium against calcium. The largest amount of calcium is transferred via the

NMDA channel. (Reviews, textbooks: Kandel 2006, Schmid, Keidel WD 1985; Unsicker 2003; Vaupel et al. 2015).

If the blood circulation stops (ischemia), the electrical charge of the cell membrane (membrane potential) collapses, since no oxygen arrives for the energy consumption of the pumps and the chemical energy reserves are used up. The available oxygen is completely consumed when the circulation stops, causing a chemical reduction of substances of the cell. In addition, harmful substances are produced by the cell (Vanden Hoek et al. 1997a). Nerve cells may die (Radovski et al. 1995).

11.2 Without oxygen, cells work against themselves: Failure of the ion pumps

Oxygen deprivation (ischemia) occurs during life, especially in the case of infarctions such as myocardial infarction or stroke. Here, the sequence of events can be followed, which can only be traced as a snapshot after certification of death and the start of cryonics. Many findings that are important for cryonics are obtained by studying infarctions or in animal experiments with oxygen deprivation. Most important is how long one can still resuscitate a patient. Therefore, there is an extensive medical knowledge about the processes involved.

A long chain of harmful events was discovered and described.

Within just two minutes without blood flow, the potassium/sodium pump now runs out of energy. Potassium ions then migrate out of the cell and sodium and chloride ions flow in. The electrical voltage, between the inside and outside of the cell (the membrane potential) finally equalizes. The cell can no longer be excited. The work of the nervous system is practically terminated

As consequence of the impaired energy production, lactic acid and other acids accumulate as waste products. During the lack of blood flow, lactic acidosis causes swelling of the inner wall cells in the vessels (Paljärvi et al. 1983). This acidification of the blood (acidosis) is further enhanced by accumulation of carbon dioxide, from which carbonic acid is formed. The pH value can drop from 7.3 to 6.7 due to acidification.

When there is a lack of fresh blood and oxygen, such as occurs in parts of tissues during a stroke or cardiac arrest, the transmitter substance glutamate is released en masse to excite nerve cells anyway. This promotes an enormous influx of calcium into the cell.

When metabolism is intense and there is sufficient oxygen, calcium activates enzymes that stimulate the formation of the energy store adenosine triphosphate in the respiratory organelles. However, if oxygen is now missing or reduced, there is an excess of electrons that form superoxide with the oxygen residues present.

The calcium elevation itself has other negative effects.

> Background information
> In this case, calcium damages the membrane of the respiratory organelles and the substance cytochrome c is released. This cytochrome c is a dangerous trigger of cell suicide (apoptosis). In addition, high amounts of calcium in the cell activate enzymes that degrade protein, called calpains (Saito et al. 1993). Even worse, high amounts of calcium activate other enzymes that degrade nuclear acids (endonucleases). These can also start the suicide of the cells.
> The cell membranes suffer the greatest damage when blood flow stops. Enzymes (some of which require calcium for their action) — release fatty substances such as arachidonic acid from the membranes (Schewe 2005). Arachidonic acid itself then damages cellular respiration. Arachidonic acid is broken down by other enzymes into other products. Inhibiting such enzymes reduces the damage that stopping blood flow normally causes (Phillis, O'Regan 2003) a sign that they are involved in the harmful chemical reactions. In addition, arachidonic acid promotes the formation of free radicals, which are also normally produced in small amounts in the respiratory chain (Cocco et al. 1999). A harmful increase in zinc also occurs in the presence of oxygen deficiency (Calderone et al. 2004).

Finally, the cell is no longer excitable, cellular respiration is impeded, harmful substances damage the sensitive membranes, among other things, and the pH is altered.

Free radicals play a major role in numerous types of damage, and their formation is caused by various reactions.

In the case of oxygen deficiency, the glycocalix of the inner wall cells in vessels is also damaged, which can lead to impaired organ function (Mathis 2021; Rabadzhieva 2012).

Dying is obviously so chemically complicated that you have to fear it for that reason alone.

> Background information
> Glycocalix is the name given to a fringe of large molecules associated with the cell membrane. The glycocalix protects the cell membrane and connects it to the surrounding components of the tissues. It also coats the inner wall of the vessels and influences the migration of molecules from the blood vessels into the tissues.

In animal studies, the damage produced by oxygen deprivation is reduced by resveratrol (Wang et al. 2002a).

Hypothermia is effective against blood-brain barrier damage in total oxygen deprivation (Preston, Webster 2004; see also Ki et al 1996; Safar 1996).

11.3 Exhausted batteries

The reduced energy reserves of the cells and the membrane damage further stimulate the release of glutamate in nerve cells. However, this only leads to overstimulation and overexcitation named excitotoxicity (Lipton 1999). There is a lack of energy to regulate the electrical voltage of the cell membrane. Figuratively speaking, the batteries for a nerve impulse are no longer charged. Glutamate now even passes directly through the membrane without the need for channels anymore when there is severe membrane damage with the formation of leaks (Phillis et al.1994), but again without generating a pulse.

But it goes even further. The loss of the chemical energy store ATP also reduces important properties of red blood cells, such as the ability to deform in order to squeeze through narrow vessels. In fact, red blood cells are larger in diameter than many capillaries. Because they are flat round discs, they fit better into vessels at an angle, but they must also be able to deform to make the passage. If this is not possible, they can form a congestion (stasis) and clog the thin capillaries already during life, so that no flow is possible anymore (see also Best B Quantifying ischemic damage for cryonics rescue. https://www.benbest.com/cryonics/IR_Damage.html; Weed et al. space 1969).

If the circulation is restored immediately, the damage is minimal and the patient or his cells can recover.

However, cryonics can usually intervene much later.

That the cells can still recover after oxygen deprivation despite all these changes is a miracle that can be proven by bringing the cells into culture (see below).

11.4 Why is human resuscitation no longer possible after 9 minutes of cardiac arrest at the latest?

Why resuscitation is no longer possible after 9 minutes, that is enormously important during dying due to total organ failure. The question when one can still intervene and how, is also decisive for the success of the cryonic suspension. We must therefore go into great detail again in the following and present the whole complexity of the processes. The changes in oxygen deficiency discussed above serve as a prerequisite for this.

11.4.1 Fatal medical resuscitation: the reperfusion syndrome

In the following, we are dealing with reactions which directly influence the methods of cryonics. In animal experiments with cooling and deprivation of oxygen, cells fall into a sleep-like state (suspended animation). Something similar has been shown in stem cells (see Latif 2012 above). Whether and to what extent cells also "switch off" at final circulatory arrest (and are thereby perhaps protected) remains to be investigated. In any case, a whole chain of reactions may take place beforehand, which is caused by the lack of oxygen.

It is important to note that reoxygenation is even more harmful than oxygen deprivation. Doctors call this phenomenon the reperfusion syndrome because it always occurs when perfusion is restarted after interrupted blood flow. That's where the doctors and the patient always have to go through if cardiac arrest is to be overcome. The problem is familiar to any physician who restarts natural or artificial perfusion after stop of blood flow or cardiac arrest (Hayashida et al. 2007).

The chains of reactions that now develop are more extensive than the initial reactions after oxygen starvation. One could get the impression that these chains become endless.

In cryonics, one question is whether to subject cells to oxygen shock without any chance of resuscitating the patient (by current means). These are cells that may already be protecting themselves by decreasing their metabolism and may no longer have sufficient metabolism to utilize oxygen or recover from an oxygen shock due to simultaneous cooling.

Thus, one should not simply mimic clinical methods in cryonics as has occasionally been done.

Reperfusion of the circulation after more than 10 min is in any case more damaging than the "lack of blood" itself. Damage to the vessels often occurs, which can lead to blockage (De Groot, Rauen 2007; Hayashida et al. 2007; Safar 1993; Solenski et al. 2002). The blockage of the blood vessels is called no-reflow phenomenon.

In a way, the lack of oxygen is now reversed. The most important reason for the occurrence of the damage is namely that the oxygen reacts, so to speak, hastily with substances in the cells and tissues that have been chemically reduced by the lack of oxygen. This results in the formation of free radicals and harmful reactive oxygen compounds in addition to all the harmful compounds already mentioned above. They play a role in most of the damage that now occurs. The oxygen, which can no longer be tamed by enzymes, shows its true colors, so to speak. With the increased formation of free radicals, the controlled energy production with the help of oxygen unfortunately lags behind. Radicals now react in their turn with the substances that have accumulated during the lack of oxygen. For example, the damage caused by reaction products of the arachidonic acid already mentioned is intensified by free radicals (Reilly et al. 1997; Zweier, Talukder 2006).

Active oxygen products such as hydrogen peroxide and superoxide (Van den Hoek et al. 1997a) increase abruptly with renewed blood flow. If a rabbit heart suffers 10 to 30 minutes of circulatory arrest, the renewed perfusion leads to a (transient) increase of such oxygen compounds within 10 to 20 seconds (Zweier et al. 1989).

After a short stop of blood flow, the renewed blood flow leads to a smaller increase in reactive oxygen products than is found during a longer stop. Thus, in the case of oxygen deficiency, these are partly already formed before the renewed perfusion (Van den Hoek et al. 1997a).

> Background information
> The inner wall cells of blood vessels contain many respiratory organelles, which form most of the superoxide during reperfusion (Di Lisa 2009; Liu et al. 2009). Enzymes of inner wall cells (NADPH oxidase and xanthine oxidase) are involved (Kahles et al. 2007; Terada et al. 1991). Superoxide then damages proteins and generates free iron, which in turn generates very harmful hydroxyl radicals. Iron fatally ensures that free radicals, which are actually very short-lived, do not disappear again quickly.
> Nitric oxide (nitrogen oxide) usually reacts positively worth mentioning. It prevents platelets from sticking together and white blood cells from sticking to the vessel wall. This nitric oxide also inhibits fat oxidation very effectively (Rubbo et al. 1994). It also promotes the formation of antioxidant enzymes that prevent harmful oxygen products and counteracts cell suicide (Dhakshinamoorthy, Porter 2004). In addition, it reduces the formation of inflammatory substances indirectly and promotes cellular connections through adhesion molecules (Mantovani et al. 1998; Rössig et al. 1999).
> However, re-circulation now turns nitric oxide into a toxin. The enzyme superoxide dismutase, which acts against free radicals, normally reduces the damage of the blood-brain barrier and the radical damage (Kim et al. 2001). It renders superoxide harmless, but nitric oxide reacts three times faster with superoxide and toxic peroxinitrite is formed in the process (Faraci 2006; Szabó et al. 2007). As a result, the inner wall cells swell (Aronovski et al. 1997) and the white blood cells stick to the vessel walls. Thus, the lumen of the vessels is constricted. Now nitric oxide is increased by several thousand times (Brown, Borutaite 2002) and it is considered to be the substance that causes the greatest damage to the inner wall cells of the capillaries in the brain under these circumstances (Gürsoy-Özdemir et al. 2000). However, one is not left completely without

an antidote. Methylene blue, which is also used as a dye, can reduce nitric oxide (Miclescu et al. 2010).

The oxygen that becomes available through re-circulation leads to accumulations of the phosphorus compound ATP as an energy store. ATP in such excess is transformed into xanthine, which is converted by an enzyme with oxygen into the harmful oxygen products superoxide and uric acid. There is also oxygen again in the respiratory organelles, and radicals and harmful active oxygen products are again formed (González-Flecha et al. 1993; Ratych et al. 1987).

Now, enzymes are also produced that break down proteins and thereby damage the blood-brain barrier (Fukuda et al. 2004; Rosell et al. 2005). Pores in the respiratory organelles remain closed during the lack of oxygen due to the acidic environment. When blood flows through again, the pH increases and this can open the pores (Halestrap 2006). This results in cell suicide or rapid cell death depending on whether energy is available for suicide (Kim et al. 2003). To make matters worse, oxygen causes membrane damage to platelets, which are responsible for clotting, and to blood cells. This happens more violently in old bodies and, in general, damage from re-oxygenation is more pronounced in old animals (Ritter et al 2008).

Many changes in reperfusion have inflammatory character. In addition to white blood cell adhesion, increased permeability of the walls of capillaries often develops, which then leads to tissue swelling. During life, activated white blood cells and inner wall cells contribute to inflammatory vascular responses and blood cell infiltration (Gourdin et al. 2009; Kuroda, Siesjö 1997). Activated white blood cells increasingly adhere to the walls of blood vessels when reperfusion is initiated (e.g., after 6-12 hours). They also produce enzymes that attack connective tissue fibers, thereby damaging the blood-brain barrier (Gidday et al. 2005; Rosenberg et al. 1998)). However, this was demonstrated in rats and mice that were still alive. In turn, the aggregation of red blood cells (like a roll of money) promotes the attachment of white blood cells (Pearson; Lipowsky 2000). Inflammatory substances, in turn, cause the inner wall cells to stick together with white blood cells when they flow through again. This effect is attenuated by cooling (e.g. hypothermia) (Ishikawa et al. 1999).

The blood-brain barrier as a specialized section of the circulatory system suffers from blood starvation and subsequent reperfusion. The permeability of the walls for molecules increases. The mentioned chemical compounds superoxide, nitric oxide, glutamate and arachidonic acid are thought to play a role in this process. Enzymes with activity against free radicals and active oxygen products counteract the increasing permeability (Armstead et al. 1992; Mayhan Didion 1996; Sage et al. 1984; Villacara et al. 1990).

Free radical damage and other membrane damage can even lead to the detachment of arteriosclerotic lesions, which are capable of producing a dangerous embolism when they flow again.

It is not surprising that reperfusion leads to increased neuronal death and, for example, increases the extent of infarction more than threefold (Aronovski et al. 1997; Li et al. 2007).

Narrowing of blood vessels is the most dangerous consequence of re-circulation in humans. The longer the blood flow has stopped, the worse the damage to the blood vessels when the circulation starts up again. Even cooling does not change this. Unfortunately, more ice crystals are formed when cooling to temperatures deep below 0°C, under the influence of free radicals and the seepage of blood fluids from the vessels. This also increases the risk of complete occlusion of vessels ("no reflow").

It is unclear whether, for example, the oxygen contained in water provides enough energy to trigger such reperfusion damage when cryoprotectant solutions flow through it. In any case, increasing cooling counteracts this. The danger is probably greater when ventilating with oxygen and performing cardiac massage during cooling.

The possibility of perfusion of blood vessels with cryoprotectant depends initially on the vascular damage that has occurred. Experiments with ink show that 60 min blood flow stop at room temperature considerably impede circulation.

Substances that inhibit blood coagulation (such as heparin and streptokinase) then also no longer improve perfusion, because an accumulation and agglutination of blood cells also occurs. This means that the vessels are already partially constricted when the new perfusion (flow) starts (Fischer, Hossman 1995) and this can exacerbate the problem.

Long ago, it was found that high flow pressure with simultaneous blood thinning partially restored perfusion (Safar 1993; Safar et al 1976). In more recent experiments, however, the benefit of this approach was not clear (de Wolf, de Wolf 2013).

The perfusion of certain parts of the brain namely the basal ganglia is particularly difficult after 15-20 minutes of hemostasis in cats. If the heart was restarted, after 30 minutes the inhibition of blood flow due to reperfusion damage was stronger than before the start of reperfusion (Fischer, Hossmann 1995).

As a result of all these different processes, in particular the lumen of the blood vessels is narrowed. The flow of blood through the circulation and resuscitation become more difficult and the inner wall cells of the vessels finally go into cell suicide (apoptosis). Ascorbic acid-2-glucoside, a vitamin C product, can be used to counteract this (Matsukawa et al. 2000).

However, the flow of cryoprotectants itself can also damage the vessels. If glycerol is added quickly, they become permeable to large molecules. Rapid removal of glycerol (with dilution), on the other hand, causes swelling of the inner wall cells and stops the flow. Thus, addition and dilution of cryoprotectant must be done carefully. If freezing is slow, the vascular cells may be constricted by ice formation in the surrounding tissue (Pollock et al. 1986).

It is hard to believe what damage is caused by a completely normal method of resuscitation. A life-saving and vital perfusion is capable of turning into a death trap. But what is even more amazing is that resuscitation is still possible. However, it matters a great deal how long the oxygen deficiency has existed.

As a result of these complicated damaging effects, resuscitation is impossible after no more than 9 minutes of circulatory arrest in humans (Aronowski et al. 1997; Cummins 1985; Herlitz et al. 2004; see also Best B Quantifying Ischemic Damage for Cryonics Rescue https://www.benbest.com/cryonics/IR_Damage.html).

In cryonics, avoiding re-circulation of blood and rapid cooling seem to be the best way to save cells before the complete cell death or at least to keep the dead cells in the tissue dressing.

However, it is of great importance that re-perfusion plays a greater role in preventing resuscitation than brain cell death (Best 2008; De Groot, Rauen 2007; Vanden Hoek et al 1996; -1997). Indeed, this also means that it is not impossible to still save brain cells.

Moreover, the reperfusion syndrome can be alleviated by antioxidative treatment or treatment with so-called radical scavengers. Free radicals play a major role in this process (Vanden Hoek et al. 1997).

11.4.2 Luck in misfortune for the cerebral cortex

Disruption of flow in the brain and ice crystal formation on subsequent cooling are lowest in the cortex of the cerebrum.

The gray cortex of the cerebrum is thankfully less susceptible to vascular occlusion than other parts of the brain. More than 50% of the blood vessels were blocked in the brain parts hypothalamus and the so-called basal ganglia of the cerebrum after 30 minutes of blood flow stoppage. Fortunately, in the gray cortex of the cerebrum it was less than 15% ().

A more pronounced circulatory disturbance is thus found in the cerebral areas below the cortex. Then follows the cerebellar cortex and finally, with the greatest damage, the medulla of the cerebellum. Reassuringly, this shows that the parts of the brain that are most important for consciousness are also the most stable (de Wolf, de Wolf 2013; Fischer, Ames 1972).

Overall, it gives some encouragement that resuscitation after heart failure is even possible.

With this, we have described the conditions that exist when cryonic preservation starts and that explain to us why cryonics is the only salvation. In the following we will describe its practice.

We will take a closer look below at the dying body as a result of the lack of oxygen and the futile repair attempts of cells, study what can still be saved from life.

Literature

Armstead WM et al (1992) Polyethylene glycol superoxide dismutase and catalase attenuate increased blood-brain barrier permeability after ischemia in piglets. Stroke 23:755-762

Aronowski J et al (1997) Reperfusion injury: demonstration of brain damage produced by reperfusion after transient focal ischemia in rats. J Cereb Blood Flow Metab 17:1048-1056

Best BP (2008) Scientific justification of cryonics practice. Rejuvenation Res 11:493-503

Brown GC, Borutaite V (2002) Nitric oxide inhibition of mitochondrial respiration and its role in cell death. Free Radic Biol Med 33:1440-1450

Calderone A et al (2004) Late calcium EDTA rescues hippocampal CA1 neurons from global ischemia-induced death. J Neurosci 24:9903-9913

Cocco T et al (1999) Arachidonic acid interaction with the mitochondrial electron transport chain promotes reactive oxygen species generation. Free Radic Biol Med 27:51-59

Cummins RO et al (1985) Survival of out-of-hospital cardiac arrest with early initiation of cardiopulmonary resuscitation. Am J Emerg Med 3:114-119

De Groot H, Rauen U (2007) Ischemia-reperfusion injury: processes in pathogenetic networks: a review. Transplant Proc 39:481-484

De Wolf A, de Wolf C (2013) Human cryopreservation research at advanced neural biosciences. In Sames KH (ed) Applied Human Cryobiology, vol. 1, ibidem, Stuttgart, pp 45-59

Dhakshinamoorthy S, Porter AG (2004) Nitric oxide-induced transcriptional up-regulation of protective genes by Nrf2 via the antioxidant response element counteracts apoptosis of neuroblastoma cells. J Biol Chem 279:20096-20107

Di Lisa F et al (2009) Mitochondria and vascular pathology. Pharmacol Rep 61:123-130

Faraci FM (2006) Reactive oxygen species: influence on cerebral vascular tone. Appl Physiol 100:739-743

Fischer EG, Ames A (1972) Studies on mechanisms of impairment of cerebral circulation following ischemia: effect of hemodilution and perfusion pressure. Stroke 3:538-542

Fischer M, Hossmann KA (1995) No-reflow after cardiac arrest. Intensive Care Med 21:132-141

Fukuda S et al (2004) Focal cerebral ischemia induces active proteases that degrade microvascular matrix. Stroke 35:998-1004

Gidday J (2005) Leukocyte-derived matrix metalloproteinase-9 mediates blood-brain barrier breakdown and is proinflammatory after transient focal cerebral ischemia. Am J Physiol Heart Circ Physiol 289:H558-H568

González-Flecha B et al (1993) Time course and mechanism of oxidative stress and tissue damage in rat liver subjected to in vivo ischemia-reperfusion. J Clin Invest 91:456-464

Gürsoy-Özdemir Y et al (2000) Role of endothelial nitric oxide generation and peroxynitrite formation in reperfusion injury after focal cerebral ischemia. Stroke 3:1974-1981

Gourdin J et al (2009) The impact of ischaemia-reperfusion on the blood vessel. Eur J Anaesthesiol 26:537-547

Halestrap AP (2006) Calcium, mitochondria and reperfusion injury: a pore way to die. Biochem Soc Trans 34:232-237

Hayashida M et al (2007) Effects of deep hypothermic circulatory arrest with retrograde cerebral perfusion on electroencephalographic bispectral index and suppression ratio. J Cardiothorac Vasc Anesth 21:61-67

Herlitz J et al (2004) Can we define patients with no chance of survival after out-of-hospital cardiac arrest? Heart 90:1114-1118

Ishikawa M et al (1999) Effects of moderate hypothermia on leukocyte-endothelium interaction in the rat pial microvasculature after transient middle cerebral artery occlusion. Stroke 30:1679-1686

Kahles T et al (2007) NADPH oxidase plays a central role in blood-brain barrier damage in experimental stroke. Stroke 38:3000-3006

Kandel (2006) In search of memory. The emergence of a new science of the mind. Siedler (Random House), Munich

Ki HY et al (1996) Brain temperature alters hydroxyl radical production during cerebral ischemia/reperfusion in rats. J Cereb Blood Flow Metab 16:100-106

Kim GW et al (2001) The cytosolic antioxidant copper/zinc superoxide dismutase attenuates blood-brain barrier disruption and oxidative cellular injury after photothrombotic cortical ischemia in mice. Neuroscience 105:1007-1018

Kim JS et al (2003) Mitochondrial permeability transition: a common pathway to necrosis and apoptosis. Biochem. Biophys. Res. Commun. 30: 463-470

Kuroda S, Siesjö BK (1997) Reperfusion damage following focal ischemia: pathophysiologia and therapeutic windows. Clin Neurosci 4:199-212

Latif M et al (2012) Skeletal muscle stem cells adopt a dormant cell state post mortem and retain regenerative capacity. Nat Commun 3:903

Li D et al (2007) Reperfusion accelerates acute neuronal death induced by simulated ischemia. Exp Neurol 206:280-287

Lipton P (1999) Ischemic cell death in brain neurons. Physiol Rev 79:1431-1568

Liu B et al (2009) Proteomic analysis of protein tyrosine nitration after ischemia reperfusion injury: mitochondria as the major target. Biochim Biophys Acta 1794:476-485

Mantovani A et al (1998) Regulation of endothelial cell function by pro- and anti-inflammatory cytokines. Transplant Proc 30:4239-4243

Mathis S et al (2021) The endothelial glycocalyx and organ preservation-from physiology to possible clinical implications for solid organ transplantation. Int J Mol Sci 22:4019

Matsukawa H et al (2000) Ascorbic acid 2-glucoside prevents sinusoidal endothelial cell apoptosis in supercooled preserved grafts in rat liver transplantation. Transplant Proc 32:313-332

Mayhan WG, Didion SP (1996) Glutamate-induced disruption of the blood-brain barrier in rats. Role of nitric oxide. Stroke 27:965-969; discussion 970

Miclescu A et al (2010) Methylene blue protects the cortical blood-brain barrier against ischemia/reperfusion-induced disruptions. Crit Care Med 38:2199-2206

Paljärvi L et al (1983) Brain lactic acidosis and ischemic cell damage: quantitative ultrastructural changes in capillaries of rat cerebral cortex. Acta Neuropathol 60:232-240

Pearson MJ, Lipowsky HH (2000) Influence of erythrocyte aggregation on leukocyte margination in postcapillary venules of rat mesentery. Am J Physiol Heart Circ Physiol 27:H1460-H1471

Phillis JW, O'Regan MH (2003) The role of phospholipases, cyclooxygenases, and lipoxygenases in cerebral ischemic/traumatic injuries. Crit Rev Neurobiol 15:61-90

Phillis JW et al (1994) Characterization of glutamate, aspartate, and GABA release from ischemic rat cerebral cortex. Brain Res Bull 34:457-466B

Pollock GA et al (1986) An isolated perfused rat mesentery model for direct observation of the vasculature during cryopreservation. Cryobiology 23:500-511

Preston E, Webster J A (2004) Two-hour window for hypothermic modulation of early events that impact delayed opening of the rat blood-brain barrier after ischemia. Acta Neuropathol 108:406-412

Rabadzhieva I (2012) Schädigung der endothelialen Glykokalyx beim Postreanimationssyndrom. Dissertation, Freiburg im Breisgau

Radovski A et al (1995) Regional prevalence and distribution of ischemic neurons in dog brains 96 hours after cardiac arrest of 0 to 20 minutes. Stroke 26:2127-2133

Ratych RE et al (1987) The primary localization of free radical generation after anoxia/reoxygenation in isolated endothelial cells. Surgery 102:122-131

Reilly MP et al (1997) Increased formation of the isoprostanes IPF2α-I and 8-epi-prostaglandin F2α in acute coronary angioplasty. Evidence for oxidant stress during coronary reperfusion in humans. Circulation 96:3314-3320

Ritter L et al (2008) Inflammatory and hemodynamic changes in the cerebral microcirculation of aged rats after global cerebral ischemia and reperfusion. Microcirculation 15:297-310

Rosell A et al (2005) A matrix metalloproteinase protein array reveals a strong relation between MMP-9 and MMP-13 with diffusion-weighted image lesion increase in human stroke. Stroke 36:1415-1420

Rosenberg G et al (1998) Matrix metalloproteinases and TIMPs are associated with blood-brain barrier opening after reperfusion in rat brain. Stroke 29:2189-2195

Rössig L et al (1999) Nitric oxide inhibits caspase-3 by S-nitrosation in vivo. J Biol Chem 274:6823-6826

Rubbo H et al (1994) Nitric oxide regulation of superoxide and peroxynitrite-dependent derivatives. J Biol Chem 269:26066-26075

Safar P (1993) Cerebral resuscitation after cardiac arrest: research initiatives and future directions. Ann Emerg Med 22:324-349

Safar P et al (1976) Amelioration of brain damage after 12 minutes cardiac arrest in dogs. Arch Neurol 33:91-95

Safar P et al (1996) Improved cerebral resuscitation from cardiac arrest in dogs with mild hypothermia plus blood flow promotion. Stroke 27:105-113

Sage JL et al (1984) Early changes in blood brain barrier permeability to small molecules after transient cerebral ischemia. Stroke 15:46-50

Saito K et al (1993) Widespread activation of calcium-activated neutral proteinase (calpain) in the brain in Alzheimer disease: a potential molecular basis for neuronal degeneration. Proc Natl Acad Sci U S A 90:2628-2632

Schewe T (2005) 15-Lipoxygenase-1: a prooxidant enzyme. Biological Chemistry vol. 383, no. 3-4, pp 365-374, https://doi.org/10.1515/BC.2002.041

Schmidt FR, Unsicker K (eds) (2003) Lehrbuch Vorklinik. Integrierte Darstellung in vier Teilen. Deutscher Ärzte-Verlag, Cologne

Solenski NJ et al (2002) Ultrastructural changes of neuronal mitochondria after transient and permanent cerebral ischemia. Stroke 33:816-824

Szabó C et al (2007) Peroxynitrite: biochemistry, pathophysiology and development of therapeutics. Nat Rev Drug Discov 6:662-680

Terada LS et al (1991) Generation of superoxide anion by brain endothelial cell xanthine oxidase. J Cell Physiol 14:191-196

Vanden Hoek TL et al (1996) Reperfusion injury on cardiac myocytes after simulated ischemia. Am J Physiol 270:H1334-1341

Vanden Hoek TL et al (1997) Mitochondrial electron transport can become a significant source of oxidative injury in cardiomyocytes. J Mol Cell Cardiol 29:2441-2450

Vanden Hoek TL et al (1997a) Significant levels of oxidants are generated by isolated cardiomyocytes during ischemia prior to reperfusion. J Mol Cell Cardiol 29:2571-2583

Vaupel P et al (eds) (2015) Anatomie, Physiologie, Pathophysiologie des Menschen. Wissenschaftliche Verlagsgesellschaft Stuttgart

Villacara A et al (1990) Arachidonic acid and cerebromicrovascular endothelial permeability. Adv Neurol 5:195-201

Wang Q et al (2002a) Resveratrol protects against global cerebral ischemic injury in gerbils. Brain Res 958:439-447

Weed RI et al (1969) Metabolic dependence of red cell deformability. J Clin Invest 48:795-809

Zweier JL (2006) The role of oxidants and free radicals in reperfusion injury. Cardiovasc Res 70:181-190

Zweier JL, Talukder MAH (1989) Measurement and characterization of postischemic free radical generation in the isolated perfused heart. J Biol Chem 264:18890-18895

12 Saving the human body: how is cryonics currently used to cryopreserve the human body?

Summary

This is now about the practical implementation of the so-called suspension (halting of life processes) of human deceased. Cryonics representatives thereby care for patients whom medicine has declared dead. The procedure is presented below with broad explanations as it is handled today in American institutes (especially Cryonics Institute, CI).

12.1 The second worst thing that can happen to us: practice of cryopreservation a saving emergency procedure

Cryonics has been called the second worst thing that can happen to us. Here we describe it.

First, cooling is needed as soon as possible, after the declaration of death.

If a living patient is in critical or terminal condition, relatives, physicians, funeral directors, and cryonics representatives in the vicinity are involved whenever possible.

Although the effects of most drugs are discussed controversially, vitamin E, fish oil, melatonin, curcumin, and N-acetyl cysteine can still be of great value if taken before organ failure-that is, during life-but less so after organ failure (Best 2008).

By having a team that watches at the bedside until the patient is deanimated (standby team), cooling can be initiated sooner after death because it is ready to go in the event of organ failure as soon as the postmortem examination leads to the pronouncement of death.

In many of Cryonics Institute's better care cases, ice was packed around the patient within half an hour. In a number of cases, this was possible earlier.

> Background information
> An ice bath cools faster than ice packs and can cool the body from 37°C to 25°C in 30 minutes. Ice bags do not cool much lower than 33°C during this time. Cooling to 10°C takes 5 hours with ice packs, 3 hours with ice cubes, and 2 hours when sprinkled with ice water. If there is flow in the blood vessels, external cooling is best applied over the head (nasal mucosa and scalp), groin and forearms, and armpit, where the skin contains many blood vessels (see also Best B Physical parameters of cooling in cryonics https://www.benbest.com/cryonics/cooling.html

A standby team can initiate circulation with a cardiopulmonary resuscitation device (thumper or Lucas for short), which rhythmically compresses the chest to cool more effectively. This involves pumping blood from the capillaries of the skin, which is cooled by the ice cubes, throughout the body. However, the blood contains waste products and is acidified.

In the event of organ failure, if a worker who is proficient in perfusion is on site in time, a "washout" (washing out of the blood) with cooled solution can be initiated. The patient is still covered with ice cubes. However, no thumper is necessary.

> Background information
> Resuscitation measures are often recommended for cryopreservation, especially ventilation with oxygen administration and cardiac massage, so that the still living cells can recover. Oxygen administration is a prerequisite for cell resuscitation. However, this does not seem to make sense cryonically, since hardly any patient is pronounced dead without failure of resuscitation measures. However, cardiac massage could promote cooling by transporting heat through the circulation. It must be said, however, that cardiac massage can be harmful as long as it moves depleted blood and unnecessary as soon as artificial circulation is present, which is actually always.
> Besides cooling, blood exchange is also performed to prevent clotting and to remove harmful substances that have accumulated in the blood during the stagnation (because surviving cells still release

waste products and the organs of detoxification, such as kidney, liver and lungs, are no longer functioning). In addition, it is possible to control the composition of the fluid flowing through the circuit and adjust the composition to adapt to cryoprotective solutions.

A cooled and stabilized patient is transported in ice to an institute, where perfusion of the circuit is initiated or, if already started, continued.

For this purpose, an artificial blood circuit is created. In principle, a solution is pumped into the vascular system via arteries and drained out of veins. The blood is gradually replaced by cryoprotectant solution.

> Background information
> But even deep-cooling without antifreeze (freezing) would not rob the patient of hope for resuscitation. The structures would be severely twisted or broken, but still present and chemically well preserved. If there are no cryoprotectants, with slow freezing, the chances of preservation would be even slightly more favorable.

Perfusion is performed in an open or closed artificial circuit. In the open circuit, the fluid drains off in the same quantity as it is pumped in and is disposed of. In the closed circuit (principle of the heart-lung machine), the fluid is returned to the vascular system. However, it can be renewed and changed; liquid can be added and drained, and the accumulation of substances in the solution can also be changed.

If you use a closed circuit, you can maintain the flow even if it is slow, so that the cells have time to absorb the cryoprotectant. You do this at least until concentrated cryoprotectant comes out off the vein.

The best results are obtained with flow at a pressure below 100 mm Hg (de Wolf, de Wolf 2013).

The flow through the head and brain is preferred, i. e. the flow is preferably through the neck arteries or other vessels (e. g. retrograde flow through the femoral artery in the direction of the arms and head) while tying off vessels that do not lead to the head.

Background information

The aorta and superior vena cava are often used. These large vessels are easy to find and cannulas can be inserted into them without difficulty. Then blood and cryoprotectant solution flow to the brain through branches of the aorta and back to the superior vena cava. If you put the cannula for the return (outflow) into the superior vena cava, you cannot determine the blood/solution for both halves of the brain separately which would be informative. But to cannulate both large neck veins (jugular veins) separately means more time and risk

The following is important for blood flow to the brain: The arteries of the spine and neck form a ring (circulus Willisii) with their branches on the base of the brain. This ensures blood flow, because if one vessel is occluded, the ring still receives blood from the other vessels. Vessels then go from the ring to the brain. In one study, the ring closure of the neck arteries at the base of the brain was complete in only 42% of participants. If the ring is not complete, for example, the posterior inferior cerebral hemisphere and part of the temporal lobe do not receive direct flow. In other studies, it was also incomplete in a high percentage. Only younger individuals and women it was more likely to find complete ring closure (Bugnicourt et al. 2009; Cucchiara et al. 2013; Krabbe-Hartkamp et al. 1998; Macchi et al. 2002; Schomer et al.1994; Tanaka, et al. 2006). Nevertheless, it seems to work that one only perfuses via the carotid arteries and does not pay special attention to the posterior cerebral arteries that come via the spine. This is the case especially when perfusing via the common carotid artery (which also supplies the face). It was found in another study that the ring was complete in 59 of 99 patients and even where it was incomplete, there was no case of insufficient perfusion when only the carotid arteries were connected to the perfusion but not the vertebral arteries (see also Best B Perfusion and diffusion in cryonics protocol lhttps://www.benbest.com/cryonics/protocol.html).

The fact that this does not lead to patient harm during heart surgery could also be due to the fact that very small arteries in the throat area can create a short circuit between the neck arteries and the cerebral arteries from the spinal column. Whether this works after organ failure has not been studied. In any case, small vessels in the cooled body are probably no longer stretchable enough to suddenly carry the blood volume of larger vessels. The previous data are not very concrete (Romero et al. 2009; Urbanski et al. 2008).

It also remains to be investigated whether the liquid really flows through the capillaries or uses bypass vessels and thus only enters the tissue to a small extent. Marking, e. g. with horseradish peroxidase and examination under the electron microscope, could also show exactly where the solution has reached. But just training an employee in these methods takes months or years, not to mention the expensive equipment and its ancillary facilities. At the same time, cryonics enjoys no support from public funds.

However, flow through the capillaries has already been observed on a thin skin of living peritoneum during cryopreservation. In this case, dextran also passed through these hair-like vessels as a large molecule (Pollock et al. 1986).

Reperfusion with blood or with cryoprotectant solution and simultaneous administration of oxygen may be beneficial within 20-30 minutes after cardiac arrest, according to many cryonics workers. After this time, however, oxygen is more harmful than beneficial.

After a short stop of blood flow, probably not too many toxic substances have been formed yet and not so many that form toxic oxidation products with oxygen.

A washout prior to transport can also be viewed critically because, like any re-flow, it can be harmful. Oxygen is a way to recover shortly after death. Simultaneous cooling and oxygen administration after pronouncement of death, on the other hand, make little sense in our opinion, which would have to be investigated.

Time and again it is found that blood replacement shortly before circulatory arrest improves artificial perfusion after organ failure. In an ink experiment, no impairment of blood flow was shown thereafter even with a circulatory arrest of up to 72 hours duration (De Wolf, De Wolf 2013)

During cooling, drugs are administered and cryoprotectant is given in increasing concentrations, at the lowest temperatures the most concentrated solutions. Cooling, after all, reduces the harmful effect of cryoprotective substances, so you can concentrate them higher at lower temperatures.

> Background information
> For most of the drugs recommended by cryonics workers today, high efficacy has not been proven. Therefore, only a small selection of the most important ones that can be given during cryonics procedures will be mentioned here. Preparations to prevent or dissolve blood clots (thrombolytic drugs) make an exception. They are indispensable. Heparin, which prevents clotting, is mainly used here.

Other agents can also dissolve blood clots that have already formed (Yepes et al. 2009).

Antioxidants (see above) can be used, but their effect is not conclusive. Further drugs are used, which stabilize the cell membranes. These include, for example, cortisone and related substances, which are also used by doctors for other reasons in many patients before death. In addition, there are drugs that protect nerve cells (neuroprotective drugs).

In most cases, the patient is cooled to 10°C as quickly as possible and the addition of antifreeze is started at 10°C at the earliest. Once the body has cooled to about 10°C, a precooled antifreeze solution is gradually added as external cooling continues. Initially at high temperatures, antifreeze agents are used in dilution. Their concentration is then increased. The aim is to enrich the antifreeze substances to about 70-75 percent, e. g. VM1 from Cryonics Institute, in order to achieve vitrification if possible.

Once about 4°C is reached during cooling, the solution becomes more viscous. It flows more slowly and often requires more time.

> Background information
> The harmfulness (toxicity) of the cryoprotective solution decreases as the temperature drops. A blood substitute solution may also contain substances with large molecules or ice blockers such as hydroxyethyl starch (HES), which prevent tissue swelling in the way that the blood protein albumin normally does (review Best B Perfusion and diffusion in cryonics protocol https://www.benbest.com/cryonics/protocol.html).
> Cryonics Institute does not yet use the option of adding substances with large molecules to the solution.

After about two hours of flow, the full concentration of cryoprotectant is reached.

Air bubbles are just as dangerous when solution is flowing through the blood circuit as they are in medicine on the living. They can block the entire flow if they become trapped in the capillaries. Because the solution is highly viscous, air bubbles are difficult to remove and must be avoided from the start.

For rapid on-site blood exchange and subsequent perfusion, the cryonics organization Alcor has a transportable miniaturized suitcase heart-lung machine. Also transportable are so-called ECMO (extracorporeal membrane oxygenation) devices.

A patient perfused with organ preservation solution can be transported to a cryonics institute or team at water ice temperature without much tissue damage if the transport time is less than half a day. However, if the transport time is too long at temperatures around 0°C, damage can be expected. For example, after more than 18 hours the damage to the circulatory system is too high to guarantee good perfusion. Of course, the initial state of the patient is also crucial.

> Background information
> Silicone oil can remain liquid down to near -100°C. Whole-body patients were cooled at Alcor in a silicone bath at a rate of 0.1°C per min. For head preparations, Alcor cooled with nitrogen gas to -135°C whereby the cooling rate has been 0.4°C per min (see also Best B Physical parameters of cooling in cryonics https://www.benbest.com/cryonics/cooling.html).

Once the final concentration of antifreeze in the solution has been reached within the blood circuit, external cooling to temperatures below 0°C can begin. Initially, dry ice is used for this purpose, which is also possible during of a transport. When the dry ice temperature of -78.5°C is reached, the patient can be stored or transported for a longer period of time. During this process in dry ice, the modern cryoprotectants remain viscous in the body.

> Background information
> However, the cells covering the heart valves were found to be damaged after 4 weeks at -80°C because recrystallization is increased at temperatures above -130°C (Feng et al. 1996). Thus, dwelling at dry ice temperature should not be too long.

A patient from Cryonics Institute or Alcor is then cooled with nitrogen vapor.

For this purpose, one uses a computer-controlled freezer with software that allows the setting of any cooling speeds.

Just above the glass transition temperature at which the tissue becomes solid, the Cryonics Institute (CI) method allows heating by 1°C and a longer pause to equalize the temperature. Then very slowly cooling to the temperature of liquid nitrogen follows, which takes about 5 days.

> Background information
> This is intended to reduce bursting and cracking (see above). However, its prevention is not yet completely possible.

Wrapped in plastic, the patient is then stored in a tank of liquid nitrogen (Best 2008).

The procedure described here was developed in part during perfusion of the kidney, which resulted in survival.

We leave the resuscitation to the future. But we have certainly stopped the harmful reactions in the body for a long time.

Below we discuss some aspects in more detail.

12.2 Prerequisites for cryoprotectant perfusion

It is quite interesting to see how each cryoprotectant and drug, as well as each method has its own advantages and disadvantages for the preservation. Of course, you cannot play around with a deceased. Therefore, one usually sticks closely to already proven methods for the cryopreservation.

12.2.1 Influence of the carrier solution during perfusion

Special organ protection solutions have been developed to keep isolated organs alive, e.g. transplant organs. These are solutions that can keep isolated organs alive (when cooled above 0°C) for hours. They contain, for example, salts, sugars and amino acids (Fahy et al. 2004). Organ protection solutions serve a large market and are accordingly carefully formulated and widely tested. They often consist of numerous components, each of which has a beneficial function.

Also in antifreeze, not only the actual cryoprotectants have an effect, but also the carrier solutions, which can themselves be cell-protecting organ protection solutions. A carrier solution also has an influence on tissue recovery (Wusteman et al. 2008).

CI's M-RPS-2 simple carrier solution consists of glucose, potassium chloride, sodium chloride diluted hydrochloric acid, and TRIS buffer. Organ preservation solutions such as MHP-2 and UW (University of Wisconsin solution) unfortunately did not allow for lifetimes of samples long enough to reach those needed for cryonics transport. Sections from the hippocampal region of the brain were used to test this. (Pichugin 2006a).

How do such mixtures work in perfusion? The solutions M-RPS-2, RPS-2 and MHP-2, which are carrier solutions for cryoprotectants, improve perfusion and reduce the formation of ice crystals. MHP-2 allowed croprotectant flow even after 48 hours of cold bloodless cooling (cold ischemia). Even 72 hours later, ice crystal formation is reduced after perfusion with MHP-2, compared to 72 hours of cooling with blood left in place (with blood causing severe flow obstructions and ice crystal formation).

On-site blood exchange (washout) in the deceased cryonics patient is thus recommended, also to remove harmful oxygen products (Vanden Hoek et al. 1997a) but depends on the composition of the solutions. None of the solutions tested (including more recent compositions by cryonics researchers) alleviated the severe tissue swelling after prolonged time in the cold, which is due to blood vessel damage. Variants of MHP-2 were generated, but did not have a more beneficial effect (De Wolf, De Wolf 2013).

At 10°C, most cells are still alive and have a very slow metabolism. Organ protection solutions have been developed to keep tissues alive at temperatures close to 0°C for extended periods of time.

Viaspan was considered a protective solution for low temperatures. It has been partially superseded by other solutions, but is still valued by users and will be discussed as an example because its composition and effects have been well studied. Viaspan is effective against cold ischemia. It can be exchanged for the blood, which after all is overloaded with harmful metabolic products after the heart has stopped.

12.2.2 Effects of substances in a cell protection solution such as Viaspan

The stated effects assume the presence of oxygen, as the solution is mostly used to enable resuscitation.

After cardiac arrest or after a transplant organ is separated from the circulation, the conversion of xanthine with oxygen into the harmful superoxide and uric acid occurs. Allopurinol inhibits this process.

Against increased oxidation and for the stabilization of the cell membranes glutathione is included. Cooling increases the permeability of the cell membranes for the glutathione itself (Vreugdenhil et al. 1991).

Membrane stability is also served by dexamethasone. Its effectiveness is also advised by physicians in cryonics.

Cell swelling due to osmotic processes can be prevented by sulfate ions because they do not enter the cell and thus remove water from the cell.

The loss of energy is counteracted by building up the energy store adenosine triphosphate. The addition of adenosine serves this purpose. It also prevents certain white blood cells from adhering to the vessel wall cells or forming aggressive oxygen products (Grisham et al. 1989). In the formation of adenosine triphosphate, the content of monobasic potassium phosphate is involved and this also counteracts acidification as well as potassium leakage from cells.

To prevent swelling of the tissue, fluid in the vessels must be held in place by substances with large molecules. Hydroxyethyl starch serves this purpose. It also reduces the adhesion of white blood cells to the vessel wall during reperfusion (Kaplan et al. 2000). Polyethylene glycol (PEG) has been used with good success to replace HES (Bessems et al. 2005; Faure et al. 2002; Franco-Gou et al. 2007).

Lactobionate and raffinose are also large molecules that do not enter the cell and prevent cell swelling. Lactobionate chelates calcium and iron and thus acts against free radicals (Marban et al. 1989). Since iron and copper contribute to the effect of free radicals, it is useful to include metal chelators for reduction of their effect (Rauen, de Groot 2004; Rauen et al. 2004a; -b; Warner et al. 2004).

The adjustment of the pH is important. HEPES buffer can be used for this purpose. It is a zwitterion buffer. It buffers even at falling temperature (Baicu, Taylor 2002), and also does not enter the cells and thus regulates the osmotic pressure.

Acidification of the blood occurs due to the use of glucose as the last energy reserve, producing lactate. Insulin serves as an antidote here.

The increased ion flux of sodium and calcium across the cell membrane during oxygen starvation is reduced by glycine (Frank A et al. 2000).

The release of damaged cell's digestive organelles can lead to protein degradation. Glutamine counteracts this and can activate so-called heat shock proteins, which protect other proteins (Bessems et al. 2005).

Anticoagulants are essential for blood circle perfusion and one can take mild anticoagulants while still alive if passing is imminent, but never without contacting the attending physician.

Changes in the carrier solution in which the cryoprotectants are dissolved were investigated, especially to reduce swelling by water absorption—edema. Addition of various salts, sugars and polymers with large molecules did not improve the situation (de Wolf, de Wolf 2013).

> Background information
> However, the vessels of cryonics patients are usually already damaged by the stagnant blood flow, especially the inner wall cells of the blood vessels (endothelial cells). Then larger molecules from the vessels can also enter the tissue and cause swelling. The company Alcor uses a carrier solution with HES and mannitol as a large molecule to maintain the osmotic pressure in the vessel.
> Lactobionate enters the cells more poorly than mannitol but is more expensive. It also draws water into the vessels. So, it can be used especially against brain swelling, if the blood-brain barrier still works. It also scavenges free radicals, especially the particularly aggressive hydroxyl radicals (Kontos 1989).

A cell protection solution is usually isotonic with the blood. It also is usually particularly suitable for the brain and kidney and has an antifreeze effect itself (Fahy et al. 1990).

12.2.3 Uptake of cryoprotectants from the blood into the cell

Above, the permeability of membranes has already been discussed. The rate at which cells take up various cryoprotectants is interesting. For isolated cells (cell cultures), the difference between the fluid inside the cells and a cryoprotective solution outside the cells is halved in 1.3 minutes (Dooly et al. 1982). For cells incorporated into a tissue, it takes longer because of the tissue components surrounding the cells (ground substances, vascular membranes, etc.). Above, it was already reported that DMSO flows through cells and tissues in seconds.

In blood cells and spermatozoa, which can be easily examined, the permeability to ethylene glycol is much higher compared to DMSO, propylene glycol (PG), and acetamide (AA) (Pedro et al. 2005).

Water like cryoprotectant passes through the cell membrane much more slowly as the temperature decreases.

Glycerol, DMSO, and ethylene glycol also decrease the rate at which water passes through the membrane (Gilmore et al. 1995).

Many studies have been performed only on cells that are easy to obtain, such as blood cells or sperm, and on oocytes, the cryopreservation of which is of particular interest. It often remains unclear whether such results can be transferred to whole organs.

In cryonics, for better absorption into the cells and to promote vitrification, solutions are usually used that have increased osmotic pressure compared to cells and tissues, and these are also not exchanged prior to freezing.

Shrinkage of the inner wall cells of vessels, especially when cryoprotectant solution is added quickly, can also have a positive effect. It can increase the receptivity of the vessels and increase the flow of solutions. This is positive as long as it does not lead to cell loss. It is even possible that blood clots may be dissolved by the cryoprotectant solution. In addition, the vessel wall may become more permeable due to damage, thus promoting the passage of a cryoprotective solution into the tissue, but unfortunately also cell and tissue swelling. Especially the breakdown of the blood-brain barrier can be favorable as long as the brain swelling is controlled by other measures. But the aquaporin water channels in this barrier are obviously better suited to allow water to cross the barrier than damage with increased permeability (Yamaji et al. 2006).

Depending on the viscosity, a flow rate of 0.5-1.5 liters per minute will be necessary to achieve a pressure of 120 mm Hg (the normal blood pressure).

Unfortunately, so far only sporadic tests have been carried out on samples of circulatory fluid and pieces of tissue from organs to see whether the concentrations of cryoprotectants really match in cells and vessels. Much more scientific research would be needed for the methods to become perfect.

If swelling occurs during perfusion of blood vessels, it can bring perfusion to a halt. Unfortunately, tissue swelling compresses blood vessels and thus hampers the flow of solution in the circulation.

A stop of blood flow may lead -as mentioned- to increased permeability of the walls of capillaries especially during renewed flow after cardiac arrest. The fluid then flows into the tissues and increases swelling there (which in turn squeezes blood vessels from the outside). Swelling can occur even during life. Inflammation is known to cause swelling and increased permeability of the vessel walls. Thus, damage can already occur, which then intensifies when blood flow finally stops.

When blood flow in the brain falls below 10-15 milliliters in 100 grams of brain, the white matter of the brain begins to absorb water, even if the blood-brain barrier is still functioning (Dzialowsky et al. 2007). Four to six hours of blood flow arrest leads to breakdown of the blood-brain barrier (Brillault et al. 2008).

> Background information
> If the blood-brain barrier still functions, the constriction of the vessels (caused by swelling) can be eliminated. For this purpose — as

mentioned – dextran can be used, which cannot leave the vessel and draws water into it (Mazzoni et al. 1990).

Mannitol has a similar effect but dissolves poorly at lower temperatures. Sodium chloride also does not cross the intact blood-brain barrier although its molecules are small. Sodium ions, chlorine ions and water, on the other hand, flow from the tissue into the cells during cell swelling and the tissue then draws these molecules from the bloodstream (across the blood-brain barrier) (Simard et al. 2007).

Substances such as mannitol with large molecules must be present outside the cells in sufficient quantities to slightly exceed the osmotic pressure in the cell. If the concentration of these non-permeating substances is low, the cells can swell twofold no matter how large the difference is between the concentrations inside and outside the cell (review Best B Perfusion and diffusion in cryonics protocol lhttps://www.benbest.com/cryonics/protocol.html).

Literature

Baicu SC, Taylor MJ (2002) Acid-base buffering in organ preservation solutions as a function of temperature: new parameters for comparing buffer capacity and efficiency. Cryobiology 45:33-48

Bessems M et al (2005) Improved machine perfusion preservation of the non-heart-beating donor rat liver using polysol: A new machine perfusion preservation solution. Liver Transplant 11:1379-1388

Best BP (2008) Scientific justification of cryonics practice. Rejuvenation Res 11:493-503

Brillault J et al (2008) Hypoxia effects on cell volume and ion uptake of cerebral microvascular endothelial cells. Amer J Cell Physiol 294:C88-C96

Bugnicourt JM et al (2009) Incomplete posterior circle of Willis: a risk factor for migraine? Headache 49:879-886

Cucchiara B et al (2013) Migraine with aura is associated with an incomplete circle of Willis: results of a prospective observational study. PLoS One 8 e71007

De Wolf A, de Wolf C (2013) Human cryopreservation research at advanced neural biosciences. In Sames KH (ed) Applied Human Cryobiology, vol. 1, ibidem, Stuttgart, pp 45-59

Dooley DC et al (1982) Glycerolization of the human neutrophil for cryopreservation: osmotic response of the cell. Exp Hematol 10:423-434

Dzialowski I et al (2007) Ischemic brain tissue water content: CT monitoring during middle cerebral artery occlusion and reperfusion in rats. Radiology 24:720-726

Fahy GM et al (1990) Cryoprotectant toxicity and cryoprotectant toxicity reduction: in search of molecular mechanisms. Cryobiology 27:247-268

Fahy GM et al (2004) Improved vitrification solution based on the predictability of vitrification solution toxicity. Cryobiology 42:22-35

Faure JP et al (2002) Polyethylene glycol reduces early and long-term cold ischemia-reperfusion and renal medulla injury. J Pharmacol Exp Ther 302:861-870

Feng XJ et al (1969) Effects of storage temperature and fetal calf serum on the endothelium of porcine aortic valves. J Thorac Cardiovasc Surg 111:218-230

Franco-Gou R et al (2007) New preservation strategies for preventing liver grafts against cold ischemia reperfusion injury. J Gastroenterol Hepatol 22:1120-1126

Frank A et al (2000) Protection by glycine against hypoxic injury of rat hepatocytes: inhibition of ion fluxes through nonspecific leaks. J Hepatol 32:58-66

Gilmore JA et al (1995) Effect of cryoprotectant solutes on water permeability of human spermatozoa. Biol Reproduct 53:985-995

Grisham et al (1989) Adenosine inhibits ischemia-reperfusion-induced leukocyte adherence and extravasation. Am J Physiol. 257:H1334-H1339

Kaplan SS et al (2000) Hydroxyethyl starch reduces leukocyte adherence and vascular injury in the newborn pig cerebral circulation after asphyxia. Stroke 31:2218-2223

Kontos HA (1989) Oxygen radicals in cerebral ischemia. Chem Biol Interact 72:229-255

Krabbe-Hartkamp MJ et al (1998) Circle of Willis: morphologic variation on three-dimensional time-of-flight MR angiograms. Radiology 207:103-111

Macchi C et al (2002) The circle of Willis in healthy older persons. J Cardiovasc Surg 43:887-890

Marban E et al (1989) Circulation. Calcium and its role in myocardial cell injury during ischemia and reperfusion. Circulation 80IV:17-22

Mazzoni MC et al (1990) Capillary narrowing in hemorrhagic shock is rectified by hyperosmotic saline-dextran reinfusion. Circ Shock 31:407-418

Pedro PB et al (2005) Permeability of mouse oocytes and embryos at various developmental stages to five cryoprotectants. J Reprod Dev 51:235-246

Pichugin Y (2006a) Problems of long-term cold storage of patients' brains for shipping to CI. The Immotalist 38:14-20

Pollock GA et al (1986) An isolated perfused rat mesentery model for direct observation of the vasculature during cryopreservation. Cryobiology 23:500-511

Rauen U, de Groot H (2004) New insights into the cellular and molecular mechanisms of cold storage injury. J Invest Med 52:299-309

Rauen U et al (2004a) Iron-induced mitochondrial permeability transition in cultured hepatocytes. J Hepatol 40:607-615

Rauen U et al (2004b) Protection against iron- and hydrogen peroxide-dependent cell injuries by a novel synthetic iron catalase mimic and its precursor, the iron-free ligand. Free Radic Biol Med 37:1369-1383

Romero JR et al (2009) Cerebral collateral circulation in carotid artery disease. Curr Cardiol Rev 5:279-288

Schomer DF et al (1994) The anatomy of the posterior communicating artery as a risk factor for ischemic cerebral infarction. N Engl J Med 330:1565-1570

Simard JM et al (2007) Brain oedema in focal ischaemia: molecular pathophysiology and theoretical implications. Lancet Neurol 6:258-268

Tanaka H et al (2006) Relationship between variations in the circle of Willis and flow rates in internal carotid and basilar arteries determined by means of magnetic resonance imaging with semiautomated lumen segmentation: reference data from 125 healthy volunteers. Am J Neuroradiol (AJNR) 27:1770-1775

Urbanski PP et al (2008) Does anatomical completeness of the circle of Willis correlate with sufficient cross-perfusion during unilateral cerebral perfusion. Eur J Cardio Thorac Surg 33:402-408

Vanden Hoek TL et al (1997a) Significant levels of oxidants are generated by isolated cardiomyocytes during ischemia prior to reperfusion. J Mol Cell Cardiol 29:2571-2583

Vreugdenhil PK et al (1991) Effect of cold storage on tissue and cellular glutathione. Cryobiology 28:143-149

Warner DS et al (2004) Oxidants, antioxidants and the ischemic brain. Exp Biol 207:3221-3223

Wustemann MC et al (2008) Vitrification of rabbit tissues with propylene glycol and trehalose. Cryobiology 56:62-71

Yamaji Y et al (2006) Cryoprotectant permeability of aquaporin-3 expressed in Xenopus oocytes. Cryobiology 53:258-267

Yepes M et al (2009) Tissue-type plasminogen activator in the ischemic brain: more than a thrombolytic. Trends Neurosci 32:48-55

13 When taken over by cryonics: condition of a medically abandoned body

Summary

Death is the absence of life. That actually says it all. But what is actually missing after the heart stops? At first, we only observe that the usual organ functions fail. But with all the damage already described, what becomes of the individual cells and the structures that life has shaped? We still find an astonishing amount of life in a corpse. How far are we from recovery in each case? Above all, how likely is recovery after thawing a human body from cryopreservation?

13.1 When is a person dead?

Cryonics waits for pronouncement of death before starting cryopreservation. In what condition is the body then? What is there still worth preserving?

Let us first look at the term "death". Death means the absence of life in a normally animate object, means loss of function and dissolution. To be dead is scientifically not a state, but the absence of a state. Death does not exist as a thing and therefore, strictly speaking, cannot have a masculine article. "The Death" does not exist.

However, it is easy to use death as a term. But then as meaning a loss of life.

13.1.1 Life is still in it

Dying cancels out life step by step. At which step does one speak of death? A "corpse" is defined by what remains of the formerly functioning, viable organization, parts of the molecular organization of a living being.

A skeleton at least still contains the individual calcium structures and can contain DNA, from which one could ideally bring a clone of the deceased into being. Life is still in it.

13.1.2 Biological definition of death and dying

We gave an overview of definitions of death earlier (Sames 2018b). One can define death as total extinction of life processes and destruction of structures. When nothing is left to tell us that it once belonged to a human being, death would certainly have been achieved in the sense of natural science. The processes (dying processes), which lead to the extinction, can last however millennia until e. g. the last bone and the last DNA are decayed. In colloquial use we would speak of dying only as long as organs show some function. Following stop of function it is decay.

As an individual, however, a human being does not only consist of his hereditary substance or special structures. I am myself only as long as my brain functions, which has developed a constant additional individuality in the course of development and during life. Consciousness and memory (Poo et al. 2016) are indispensable for being human. Their final loss can be taken as the final loss of individual human existence.

In practice, medicine needs a definition of death that allows rapid certification of death and early burial or organ removal.

Such a practical definition describes the stage reached by dying/decay rather than the final death, the total absence of life. But it presupposes the final loss of consciousness and of the possibility of revival, that is, the loss of individual existence.

Characteristics include: loss of organ function, failure of resuscitation, and initial signs of decay.

A well-known example is cardiac arrest, which used to be considered "death," but today is only one sign of death among others. Today, it is often possible to restart the heart.

After 5 to (at most) 9 minutes stop of blood flow in the brain (e. g., due to blood loss or cardiac arrest), a person cannot be resuscitated today, as discussed above.

Death is considered certain when the electroencephalogram (EEG) no longer reveals brain waves and the heart has stopped, resuscitation (resuscitation) has been attempted unsuccessfully, and cerebral blood flow has been interrupted for 10 minutes or more. Cadaveric lividity (livor mortis) or rigor mortis are examples of certain signs of death, one or more of which must be demonstrated for death to be established, especially if an electroencephalogram (EEG) is not done to detect brain waves. Lividity appears no earlier than after 20 minutes of circulatory arrest (delayed if refrigerated), and rigor mortis appears later.

Since "death" must be waited for the start of cooling the medical determination and certification of death plays a major role for cryonics. It has undergone several transformations over the ages and could be comprehensively changed anew by cryonics. Today, a human being is also declared dead because one refrains from developing further measures to resuscitate or preserve him. However, this is true only if such measures are conceivable. With the advent of cryonics, they are.

13.1.3 Evidence of death

1. Until about the 1950s, it was assumed that death was irreversible as soon as the heart stopped beating. If resuscitation is finally abandoned, this means the end of life. To be sure, however, it must be determined whether resuscitation would still be possible.
2. evidence of definite signs of death, in particular livor mortis and rigor mortis, is therefore a legally binding requirement for the certification of death after cardiac arrest at the medical post-mortem examination in Germany.
3. it is widely believed that after blood flow stops for 5-9 minutes, the brain is dead.
   If blood flow is artificially maintained, "brain death" must be proven by a complicated brain death diagnosis and an observation period must be observed, which depends on the type of brain damage. The death of all brain parts must be proven. Instrument based controls supplement the diagnostics.

4. it is discussed whether a human being is dead, if only the lower brain centers survive, but not the cerebrum in which one assumes the consciousness. There is also debate about minimal life processes in the brain that are difficult to detect. A conflict may arise in organ removal between preservation of the patient's life (when this is unclear) and preservation of a transplant organ. It can usually be avoided.
5. In the case of sudden cooling (drowning in icy water), death may be greatly delayed, i. e. resuscitation may still be possible after 1 hour or longer.
6. pathologists have pointed out early — mainly because of the long preservation of the organ and body structures — the gradual development of death (more precisely the gradual dying/decay), about which we now have more precise ideas. When the brain is finally dead, remains an open question, because death is a process. It does not occur as a sudden and irrevocable event. What can occur suddenly is the failure of organs to work.

13.2 More than dead? How dead is finally dead?

One can determine that an individual is dead if his individual consciousness and memories cannot be recovered.

However, one must realize that the lack of supply and the death of the cells only begin with the cardiac arrest, while the body lies motionless and without the usual function of the large organs. As is known, it also becomes cold when the organs no longer produce heat, when the blood does not transport heat.

The changes discussed above due to oxygen deficiency then take place in the cells.

13.2.1 A small difference

For cryonics, the ultimate question is how long after "death" cryonics makes sense. It can take 2 hours or longer until death is pronounced in Germany.

Some animal species endure the lack of oxygen a little longer than humans. However, in one study of mice that suffered 3 minutes of blood starvation, only 46% survived for at least 72 hours, whereas 40% died by then (Menzebach et al. 2010).

Brain cells of rats recovered after 10 min of oxygen deprivation when given oxygen. Animals also survive the 9 min limit at room temperature. Most cells are still alive after 22 min of circulatory arrest. However, a small number of cells are dead (Pichugin et al. 2006a; Safar 1993).

Monkey brains (rhesus monkeys) subjected to a one-hour blood flow arrest (by vascular occlusion) at normal temperature and not reperfused showed a failure of overall electrical activity in the brain (with the exception of individual animals). Certain recovery signs persisted both electro physiologically and metabolically for at least one day (Bodsch et al. 1986; Hossmann et al. 1970; -1973; -1975; -1986; -1998; Hossmann, Zimmermann 1974).

Failure of resuscitation, as discussed, depends initially on obliteration of the blood vessels. In the first and second hour, the most important consequence of the lack of energy is constriction of the blood vessels (Hossmann 1988; Hossmann, Zimmermann 1974; Pearson et al. 1977; Ratych et al. 1987; Wu et al. 1998).

In organs that can be transplanted, the duration of blood flow arrest plays a critical role in the outcome of transplantation (Opelz 1998).

On the other hand, the medium-sized vessels that regulate resistance to the heart's blood pressure slacken and are thus dilated, which can cause the blood pressure to drop. This can be counteracted by drugs that cause vasoconstriction, e. g. epinephrine. To overcome the constriction of other vessels, blood pressure can also be increased, as was done in an experiment in which in addition heparin was used to prevent clotting. Insulin was also given and acidity was regulated. This allowed resuscitation after more than the critical time (Safar et al. 1976; Shaffner et al. 1999).

Even if resuscitation fails, there is no evidence that all neurons of the brain are dead after 5-9 minutes of oxygen deprivation. Failure to resuscitate does not equal total brain death. This is also true when cells do not function as usual to form electrical excitations. For a long time, physicians have passed over this small difference the difference between pure loss of function and loss of cells and other components of biological tissues. However, it is an important question for cryonics whether the brain already shows general cell death and decay of structures after 10 minutes of circulatory arrest.

It is interesting to know how cells die when deprived of oxygen in order to determine the type of cell death (Lipton 1999).

13.2.2 How do cells die?

We know two forms of cell death.

Necrosis is the name given to rapid cell death. It does not require energy. It is not active and is not controlled by the DNA program.

Apoptosis, the active suicide of the cell, on the other hand, is genetically programmed by the DNA. It also requires energy and time.

The slowed cell death that follows a deprivation of oxygen (ischemia) is an active process. It shows characteristics of cell suicide (apoptosis). For example, the gene Bax, which promotes cell suicide, is initiated prior to cell death upon circulatory arrest (Chen et al. 1996; Chu et al. 2002; Yakovlev, Faden 2004).

Excitotoxicity, the overexcitation of neurons in the absence of oxygen, can also cause cell death (necrosis) (Choi 1996).

Rats showed signs that the suicide program was running in the cells after only 15 minutes of blood flow stop and subsequent re-circulation for 8 hours. For example, the enzyme caspase-3 (acting in the apoptosis pathway) was increased in many brain regions, in the hippocampus region even after 8-24 hours by 10-fold (Chen et al. 1998).

The fact that the human brain cannot be revived after 5-9 minutes could in part be due to the fact that the cells which do not die in the initial phase quickly enter the suicide program. However, its termination can take a long time. Only then the cells would be dead.

In human stroke, it has been shown that cell suicide starts in the first two days. In this process, the destructive enzyme caspase 3 is increased. But fortunately, cell suicide is still not completed. The appearance of the dying cells is more similar to necrosis than to cell suicide. The worst process in cell suicide, the cleavage of DNA develops late (Love et al. 2000). However, in stroke, the brain is not affected as a whole.

Other organs and cells have also been shown to have a slow onset or incomplete progression of the cells' suicide program under various forms of circulatory arrest.

> Background information
> The cells behave completely differently in different tissues. Thus, in the light-sensitive retinal cells of the rat, suicide processes were found after only 90 minutes of blood starvation. Cell suicide is promoted in the eye by light energy. Fractures of the genetic substance DNA also occurred here only after other microscopic changes were already seen.
> A stop of circulation in the neighboring pigment epithelium of the eye showed completely different symptoms. In contrast to the retina, the pigment epithelium contains only few nerve cells. Suicidal phenomena occurred there only many hours later than in the retina (Hafezi et al. 1997).
> In the removed human kidney, for example, during 85 minutes without blood at 37°C, the substances Bax and caspase-9 increased. These substances promote the suicide of the cells with the help of the respiratory organelles (mitochondria). However, there are also proteins which inhibit the suicide process such as Bcl2 and cFLIP. They decreased, but the pathway to the particularly dangerous caspase-3 was not opened, not even by a special pathway via the so-called death receptor. Suicide remained incomplete (Wolfs et al. 2005).

So, the suicide of the cells occurs in different ways during stop of circulation and it can proceed incompletely, in dead, it probably must. In case of final heart failure, it may be still at the very beginning (even eight hours after the final organ failure) at least in the nervous system and in the kidney.

The suicide of our cells seems to serve life when healthy, namely by eliminating damaged cells. Therefore, it does not benefit the deceased. Here all cells start to be damaged.

13.2.3 The sensational survival of nerve cells

It is known that mesenchyme type stem cells from mammals, including humans, can be grown long after organ failure and divide up to 30 times or more in cell culture (e.g., Bliss LA et al. 2017; Cieśla J, Tomsia M 2021; Lee K et al. 2017; Saito T et al. 2020; Sames 1980).

It has now been shown that in the brain, neurons and other cells continue to show signs of life even hours after organ failure ("death").

Living and growing stem cells can be obtained, for example, from the forebrain of rats up to 6 days after circulatory arrest. Nerve stem cells are also present in human brains, spinal cords and retinas for up to one day after organ failure. They make up to 70 divisions (passages) when kept in cell culture. Stem cells can develop (differentiate) into mature neurons (Fuchs et al. 2000: Klassen et al. 2004; Lovell et al. 2006; Mansilla et al. 2013; Palmer et al. 2021; Schwartz et al. 2003; Verwer et al 2002a; Xu et al. 2003).

Dormant stem cells are also present here and can be activated in the event of brain damage (Llorens-Bobadilla et al. 2015; Mansilla et al. 2013).

In addition, neurons can be formed from astroglial cells-which are closely associated with neurons (Berninger et al. 2007; Fuchs et al. 2000; Heinrich et al. 2011; Sanal 2004).

13.2.4 Resurrection of the pigs — survival of animal nerve cells

As mentioned, some animals (and thus their brain cells) survive longer periods without blood movement at room temperature than do humans (Safar 1993).

Inner ear tissues (including sensory cells i.e. neurons) of newborn mice still show preserved tissue structure and structure of cells 10 days after organ failure without significant change in the number of cells. After that, the number of dead cells reaches 1/3 to 1/2 of the number of living ones. Stem cells can also still be isolated (Senn et al. 2007).

Even after 1-6 hours of cardiac arrest and reperfusion, excitation is still found at neural connections in rodents (Charpak,

Audinat 1998). In one experiment, rats were left at room temperature (i.e., without cooling) for various times after heart arrest before sections were taken from the hippocampal part of the brain. After 0.6-0.7 hours, electrical activity had dropped to 50%, and after 3 hours it was about 25% (Leonhard et al. 1991). An almost identical experiment was reported by Pichugin (2006a). He used the potassium/sodium ratio as evidence of survival and his rats were younger. After 3 hours, the detection still indicated 50% survival. He found the ratios much more favorable when the brains were kept in the cold at 2-4°C after rapid isolation (so-called cold ischemia). Here, a survival of 50% up to 10 hours was demonstrated. After 50 hours it was still 20-30%. After about 4 hours, however, he found no significant impairment of survival. Thus, with complete organ failure and absence of oxygen, survival of a high percentage of cells can be expected. Using Pichugin's method, neurons cannot be distinguished from other cells in the brain. Moreover, it is impossible to say whether the drop is due to the death of a number of cells and the preservation of others. It could also be that the pumps in all cells are slowly failing at the same time. So further investigation would be needed here.

In the other investigations, a total value of electrical activity was tested, which cannot access individual cells. Thus, the exact number of surviving neurons remains to be determined.

> Background information
> Rat tissues showed the first signs of dissolution at room temperature only 9 hours after organ failure. Ribosomes, the corpuscles of program-controlled protein formation disappeared from the cells. 2.5 percent of the neurons contained caspase 3 which occurs in cell suicide (Sheleg et al. 2008). An examination of the ultrastructure of rat brain neurons was performed using an electron microscope. It showed that after three hours of circulatory arrest, the respiratory organelles were only slightly swollen. After 5 hours, the swelling was more severe. After 24 hours, the ruptured respiratory organelles could still be distinguished. However, if after 3 hours of circulatory arrest the circulation was reperfused for 2 hours, the cells were severely swollen and the respiratory organelles were destroyed. If circulation was continued for up to 24 hours, the

respiratory organelles disintegrated (Solenski 2002). Cell death resembled necrosis (Jones PA 2004).

There are animal experiments with vascular occlusion. They are mainly used to study stroke: In the occlusion of the middle cerebral artery of rats, watery brain swelling (edema) occurs in the first 3 hours when a renewed blood flow is initiated (Slivka et al. 1995), whereby the blood brain barrier becomes permeable to sodium, which leaves the vessel and draws water into the brain tissue. Only after two days has the barrier become permeable enough for protein molecules (i.e., larger molecules) — such as albumin present in the blood — to pass through it as well. The throttling of blood flow leads to a zone of oxygen deficiency (Dawson et al. 1997). In such an experiment, a relatively small amount of dead neurons (15% of the population) was found before 5-6 hours. In the non-perfused zone, only single cells were undergoing suicide (as demonstrated by the TUNEL reaction). Constriction of the artery was by an inserted suture. Cells and substances may migrate from the surrounding zone, which is damaged but still alive (it is called penumbra). How far migration of substances from this zone can supply the cells in the damaged zone is difficult to judge. Even after 3 to 4 days, single intact neurons have been found in the decaying (necrotic) district of the damaged cortex (Garcia et al. 1994; -1995; Gotoh et al. 1985; Pulsinelli 1982; Rupalla et al. 1998; Takahashi, Macdonald 2004).

Cerebral neuron type cells were taken from the cortex of the rat frontal brain. After removal, the brains were stored at 22°C or at 4°C and sections were taken after different times. The number of living cells was preserved four times longer at 4°C than at 22°C. Aimed for were 20% of living cells at each of the two temperatures. The tissue contained 160 000 cells per milligram. Of the cells collected immediately after death, 9% of the neurons were viable. Surviving, isolated neuronal cells accounted for 0.5-2.7% of the cells in the brain. After 5 days in culture, 23-42% of the originally isolated neurons were alive. (Viel et al. 2001). In this experiment, the tissue was disintegrated and no reliable conclusion can be drawn about the survival of cells in the uninjured tissue. Nevertheless, a large number of living cells could still be isolated 24 hours after killing.

Brain size is not an absolute barrier to cell survival, as results in larger animals show. But human brains belong to a different size and stage of development.

In cats, even after two hours of blood flow arrest at normal temperature, the membranes of the digestive organelles (lysosomes) were intact. This is very reassuring because otherwise there would be leakage of degradative enzymes and self-digestion (Kalimo et al. 1977; Yamashima, Oikawa 2009). Cat brains developed brain waves again after one hour of oxygen deprivation and subsequent supply, and one cat could be resuscitated (Hossmann et al. 1970; Grosse Ophof et al. 1987).

Dogs show significant damage after only 10 minutes of circulatory arrest (blood pressure 0mmHg). They do survive, but if examined 96 hours later after re-perfusion (which certainly leads to increased impairment), considerable damage is seen, especially in the hippocampal region of the brain. Approximately 25-100% dead cells occur depending on the brain region. In the temporal region and the islet of the cerebrum one finds - 25-50% of dead cells depending on the region. 20 min of circulatory arrest is survived with increased damage (Radovsky 1995).

In a study by White et al. (1966), dog brains kept at about 2°C for up to 4 hours regained their electrical activity when rewarmed. (see also Best B Quantifying ischemic damage for cryonics rescue. https://www.benbest.com/cryonics/IR_Damage.html)

Brains of pigs (6-8 months old), which were slaughtered in the usual way (exsanguination and decapitation after electric shock), could be kept alive up to 4 hours after collection when perfused with a special solution ("BrainEx") at 20°C, allowing oxygenation. Measured were, energy balance, cell death and tissue architecture. Upon stimulation, single electrical excitations occurred. Overall excitation in the EEG was not derived (due to constraints imposed by an ethics committee), but was blocked, limiting the significance (Vrselia et al. 2019), whereas in the above experiments by Leonhard et al. after hours of oxygen deprivation of rat brains, clear network excitation was detected in slices, and in the experiment by White, electrical excitation was also detected. The pig experiments say nothing about the situation in humans with organ failure. The investigators cite the studies of Verwer et al. (see below), so they have human brain reanimation in mind.

Washout, transport, preparation of the brain and connection to the dedicated perfusion unit took 4 hours. Then perfusion started with BrainEX and, in control brains, with saline, which, however, induced a "no reflow" phenomenon after 6 hours.

The procedure is similar to that of Safar and Tisherman, which, however, does not involve killing but immediate rescue by perfusion in animal experiments. In clinical cases with reanimation, the normal procedure also includes an early interruption of heart arrest and oxygen deprivation.

The international attention that the pig's head results attracted is incomprehensible in view of the earlier experiments cited above. The news went through the tabloid press as a sensation. Members of leading German institutes as well as the Ethics Council felt compelled to take a stand. The definition of brain death (crucial for the removal of transplant organs) seemed to be in question (published at SMC).

One explanation for the uproar would be that this was not about killing for scientific purposes, but on the contrary a chance to avoid such experiments. The reawakening of the poor massacred and already processed creature is emotionally touching. But ultimately, it is simply about a lack of information. The earlier scientific experiments on animals are not so popular, but their results can be as sensational as the "resurrection" of pigs and they should be available to the public in the same way. The media also have a duty in this respect.

It should be noted here that the experiments of Safar, Tischerman and others described above, the results of Pichugin unfortunately not published in a leading journal, and the very early results of White and others were hardly mentioned in this context. Also, the reversible absence of excitation in the EEG during hypothermia and cardiac arrest in surgery shows that a brain can recover from oxygen deprivation (but after shorter duration).

On the whole, previous animal studies indicate that after 4 hours of oxygen deprivation, either the damage is still minor or a large proportion of the cells are still viable, regardless of the size of the brain. In this respect, the pig experiments do not bring any gain in knowledge as long as the full function of the brain is not restored, which is unfortunately still unthinkable with the available methods on the human brain.

The procedure is not to be despised as an additional, inexpensive test model for studies on the brain. Above all, it opens up the possibility of saving animal experiments and shows the necessity of deciding on animal welfare ethical questions; especially about the question how cruel or not it is to let a conscious brain exist cut off from all information and reaction possibilities in the "dark". The earlier experiments show that slices of the brains are perfectly suitable for investigations, so that one can work below the level of consciousness, as far as this can be judged today.

13.2.5 Human nerve cells

Of course, it is particularly interesting to ask whether there are still living cells in the human brain, which fails so quickly when oxygen is depleted.

In neurons from human brain tissue taken 3-6 hours after total organ failure, it was indeed shown that oxygen metabolism and the so-called axonal transport — i.e. the transport of substances in the large processes of the neurons — recovered under suitable culture conditions (Dai et al. 1998).

Brain cells were isolated from human brain slices taken at mean of 2.6 hours after pronouncement of death. Over 82% percent of the isolated cells were alive. However, only the portion of the processed brain tissue that contained the most cells was examined and there were a lot of decay products perhaps from decomposed cells (Konishi et al. 2002).

In sections from the cerebral cortex taken 2-8 h (mean 4.2 h) after death, brain cells could be kept alive in culture. These brains were even exposed to room temperature for 1-2 h during transport (refrigeration would probably have been more favorable for survival). 30-50% of the cells were alive 20-30% were damaged but not dead. The slides remained viable for at least six weeks and maintained tissue structure and diversity of (neural) cell types. Cells recovered from the agonal state in vitro and viability declined after 2-3 weeks. Field potentials were recorded by multiple electrodes. Hyaluronan is produced (mainly by astrocytes) indicating viability of the cells. Inside the slides macrophages derived from microglia are involved in repair processes. One of the most exciting discoveries was that the dead cells appeared to remain in their normal position for a long time. This could perhaps one day be favorable for tissue reconstruction. In tissue cultures, the cells lived for weeks and could be used in experiments to analyze Alzheimer's disease (Plug et al. 2024; Qui et al 2018; Verwer et al. 2002a; -2002b; -2003).

The rapid cell death, the necrosis runs without energy and therefore completely. However, the energy for the scavenger cells from the blood (macrophages) as well as for the microglial cells located in the brain (see Lourbopulos et al. 2015) seems to be missing here. Thus, the dead cells are not eaten (phagocytosed). They remain in their location in the tissue association.

Nevertheless, similar findings have been made in humans and rats, according to which they develop signs of cell suicide early after circulatory arrest. In this context, the suicide enzyme caspase 3 is elevated in many brain regions (Chen et al. 1998; Pearson et al. 1977).

With the number of cells that could be resuscitated in culture, one must assume that suicide is not very advanced when the cells stop their metabolism (even after 8 hours of cardiac arrest). However, at this stage they are held in cryopreservation and as far as we know they can remain in liquid nitrogen as they are brought in, i. e. in their organs and with the damage caused by the dying process and cryopreservation. Only when warmed up they can resume life or complete their suicide — theoretically still after millions of years.

13.2.6 A suicide that lasts millions of years

Only a small fraction of the cells die rapid cell death (necrosis) after blood flow stops, as shown by the examples above. The vast majority apparently enter the chemically controlled suicide program of apoptosis, so that in dogs, for example, they are finally dead only after hours (Radovsky et al. 1995).

The studies carried out so far—especially for humans—do not yet provide reliable figures for the survival of nerve cells in the absence of oxygen. However, if 50% survive, this also means that 50% are dead.

In this context, delayed cell death in the absence of oxygen is an active process and shows at least some signs of cell suicide.

Overall, however, the findings appear so favorable that one wonders whether it would not be quite easy to revive a brain after hours of cardiac arrest if the methods were vigorously developed further and the stem cells activated. Here, however, the obstruction of the vessels is still an insurmountable obstacle, because the tissue would at least have to receive blood for regeneration.

Resuscitation of patients drowned in ice water shows that no critical cell loss occurs with immediate cooling.

For cryonics, the hurdles remain: Harmfulness of antifreeze, crystallization and cracking.

We are not yet able to describe memory, consciousness and individuality completely or to say what would be necessary for their preservation. Of course, complicated theories exist about it. For cryonics, it is to be hoped that memory stores survive cryopreservation. It would be a disaster for cryonics if the memory stores were so unstable that they were lost in the absence of oxygen or damaged by cryogenic temperatures. Nerve cells and other cells related to them experience changes during long time memory storage. A high number of processes involving many substances in cells and cellular surroundings, in cellular connections by synapses and cellular contacts seem to be related to memory longtime storage functions and so are formations of activated nerve cell groups. (see, e.g., Costa-Mattioli et al. 2009; Josselyn et al. 2015; Lisman 2015; Ortega-de San Luis, Ryan 2022; Poo et al. 2016; Richter, Klann 2009).

We hope, the many interconnected changes stabilizing memory make engrams resistant to oxygen loss and cooling to cryogenic temperatures. In favor of its stability, we mention the lifelong preservation of memory contents and the fact that patients fully recover after suspended cardiac action. In 2015, Blackiston et al. indicated that memory content even survives total nervous system remodeling during, for example, insect metamorphosis.

There are even some hints to the ability of ganglia and brains to outlive freezing temperatures. We have already mentioned mammals reaching freezing temperatures of their bodies in winter and there are insects and nematodes outliving cryogenic temperatures and — as we may well assume — without loss of the behavior of their species. Hamsters endure freezing of 60 % of their brain water, but the brains were not cooled down to cryogenic temperatures. Their behavior was not affected by freezing. A direct proof of a memorized reaction has been demonstrated following cryopreservation of the nematode Caenorhabditis elegans (s. 15.1).

Ashwin de Wolf (2021) has investigated how long the morphological preservation of the brain makes cryopreservation useful.

13.2.7 Not only nerve cells are masters of survival

We do not know what possibilities of recovery will still come through future development. Therefore, it is reasonable to be generous with the time after death at which we consider cryonics to be useful, and to continue to subject people to cryonics for a long time after organ failure. In practice, severe delays occur in most cases, but one does not have to give up because of this.

What happens in a dying body that cannot be resuscitated is sparcely studied and probably of little interest to normal medicine.

Ironically, the only way to prove that such a body and brain are still viable would be to resuscitate them. But this is still impossible today.

Cryonic science is needed here, which could also help to further delay the time of death and better preserve transplant organs. Generous funding of this science would be of extraordinary benefit, its omission hardly justifiable.

However, many cells of different organs outlive oxygen deprivation and even pronouncement of death. It is known that stem cells from connective tissues of mammals, including humans, — as mentioned above — can be grown long after organ failure and divide up to 30 times or more in cell culture (e.g., Sames 1980).

The last legion of defense cells

It has been speculated that oxygen deficiency, hyperacidity and lack of nutrients, as well as other deficiencies due to organ failure, are what really get stem cells going — unlike other cells — and weed out the less robust among them. The cells of different organs are subjected to a real toughness training. In this way, cells are assembled that are more robust and more successful than others. They do not respect the death of the body. Possibly they serve a last effort to repair tissues despite general damage, like defenders of a city repairing a wall prism under enemy fire.

If the repair attempt fails, the cells fall into a deep sleep, so to speak, to preserve their strength. A substance known as a cytokine called interferon gamma is involved in this process. The cells begin an orderly retreat, so to speak (Mansilla et al. 2013).

We mentioned the survival of brain cells and brain stem cells and there for example are also other stem cell and cartilage cells (Alibegovic 2014), corneal endothelial cells, cells of the trabecular meshwork of the eye (Sames 1980), cells of the eye lens (Mayer, Sames 1981) proven by cell cultures to be alive following stop of circulation and pronouncement of death.

And here's the kicker: muscle stem cells from humans can retain their regenerative power up to age 90 and up to 17 days after organ failure (Latif et al. 2012)

Lately Andrijevic et al. (2022) reported on restoration of cell and tissue functions following one hour of heart stop in pigs. The authors used an advanced perfusion system to overcome the obstacles normally prohibiting blood circulation, caused by oxygen deprivation, to revive many cell types in different organs. Thereby they opened a chance to restore function of organs and — hopefully — the organism. They have been able to demonstrate a number of vital functions in the cells of different organs. Thus, dying of cells post mortem needs more time as previously estimated and reanimation may in principle be possible even hours after the stop of circulation.

13.3 The new death

Overall, however, a new conception of "death" becomes visible, a negative one: as long as we can assume that the brain can be repaired, it is not dead.

In 2014, in a lecture (see Sames 2018b), we summarized the state of the science on brain survival (slightly supplemented here) in the following sense:

- Cells that complete suicide normally disintegrate and the debris is taken up by phagocytes. However, it does not look like all brain cells go through suicide and are eaten at the same time in final organ failure. The brain does not look as if all neurons have perished even hours after the heart has stopped.
- Even without cryoprotectant, 80% of nerve cell connections (synapses) retain the metabolic properties of cell connections taken out alive after cooling to -70°C (Hardy et al. 1983).
- The cell structure of the brain is preserved for a long time, because the dead cells (as a result of the lack of energy) are not degraded.
- Different types of cells can replace nerve cells.
- Stem cells endure oxygen deprivation.
- A high redundancy of structures and information can compensate for memory losses.
- The plasticity of the nerve cells uses the redundancies.
- Even after 8 hours of circulatory arrest, many neurons (estimated up to 50%) are still viable in the human brain
- In brains of animals totally deprived of oxygen, corresponding survival rates of cells were found.
- Suspended Animation allows for hours of suspended brain function.

- Several studies on different animals show that up to about 4 hours of absolute oxygen deprivation, activities of neurons can be recovered.

By exploiting, developing, and medically using these facts, one could attempt to reanimate the brain even after prolonged organ failure.

However, the assumption about an availability of stem cells in the adult brain was shaken by a publication of Sorrells et al. (2018), according to which the spontaneous formation of new neurons is practically detectable only in childhood in the brain part hippocampus. The results have yet to be confirmed and it remains to be investigated whether it follows that there are no cells left in later life that can replace neurons, which is quite unlikely.

Literature

Alibegović A et al (2014) Viability of human articular chondrocytes harvested postmortem: changes with time and temperature of in vitro culture conditions. J Forensic Sci 59:522-528

Andrijevic D et al (2022) Cellular recovery after prolonged warm ischaemia of the whole body. Nature 608: 405-412

Berninger B et al (2007) Functional properties of neurons derived from in vitro reprogrammed postnatal astroglia. J Neurosci 27:8654-8664

Blackiston DJ et al (2015) The stability of memories during brain remodeling: A perspective. Commun Integr Biol 8e1073424

Bliss LA et al (2012) Use of Postmortem Human Dura Mater and Scalp for Deriving Human Fibroblast Cultures. PLoS One. 7(9):e45282. doi: 10.1371/journal.pone.0045282

Bodsch W et al (1986) Recovery of monkey brain after prolonged ischemia. II. protein synthesis and morphological alterations. J Cereb Blood Flow Metab 6:22-33

Brown GC, Borutaite V (2002) Nitric oxide inhibition of mitochondrial respiration and its role in cell death. Free Radic Biol Med 33:1440-1450

Charpak S, Audinat E (1998) Cardiac arrest in rodents: maximal duration J Biosci compatible with a recovery of neuronal activity. Pub Nat Acad Sci USA 95:4748-4753

Chen J et al (1996) Expression of the apoptosis-effector gene Bax, is up-regulated in vulnerable hippocampal CA1. J Neurochem 67:64-71

Chen J et al (1998) Induction of caspase-3-like protease may mediate delayed neuronal death in the hippocampus after transient cerebral ischemia. J Neurosci 1:4914-4928

Choi D (1996) Ischemia induced neuronal apoptosis. Curr Opin Neurobiol 6:667-672

Chu D et al (2002) Delayed cell death signaling in traumatized central nervous system: hypoxia. Neurochem Res 27:97-106

Cieśla J, Tomsia M (2021) Cadaveric Stem Cells: Their Research Potential and Limitations. Front Genet 12: 798161. doi: 10.3389/fgene.2021.798161

Costa-Mattioli M et al (2009) Translational regulatory mechanisms in synaptic plasticity and memory storage. Prog Mol Biol Transl 90:293-311

Dai J et al (1998) Recovery of axonal transport in "dead neurons". Lancet 351:499-500

Dawson DA et al (1997) Temporal impairment of microcirculatory perfusion following focal cerebral ischemia in the spontaneously hypertensive rat. Brain Res 749:200-208

De Wolf A (2021) Improving cryonics case outcomes. Biostasis the annual biostasis conference, Zurich

Fuchs E et al (2000) In-vivo-Regulation adulter Neurogenese. Wie groß ist das Differenzierungspotential multipotenter neuronaler Stammzellen? In Sames, K (ed) Medizinische Regeneration und Tissue Engineering. Ecomed, Landsberg, V-9, pp 1-8

Garcia JH et al (1994) Brain microvessels: factors altering their patency after the occlusion of a middle cerebral artery (Wistar rat). Am J Pathol 145:728-740

Garcia JH et al (1995) Neuronal necrosis after middle cerebral artery occlusion in Wistar rats progresses at different time intervals in the caudoputamen and the cortex. Stroke 26:636B-642B

Gotoh O et al (1985) Ischemic brain edema following occlusion of the middle cerebral artery in the rat. I: the time courses of the brain water, sodium and potassium contents and blood-brain barrier permeability to 1251-albumin. Stroke 16:101-109

Grosse Ophoff B et al (1987) Recovery of integrative central nervous function after one hour global cerebro-circulatory arrest in normothermic cat. J Neurol Sci 77:305-320

Hafezi F et al (2009) Retinal degeneration, apoptosis and the c-fos gene. Neuroophthalmology 20:143-148

Hardy JA et al (1983) Metabolically active synaptosomes can be prepared from frozen rat and human brain. J Neurochem 40:608-614

Heinrich C et al (2011) Generation of subtype-specific neurons from postnatal astroglia of the mouse cerebral cortex. Nature Protocols 6:214-228

Hossmann KA (1988) Resuscitation potentials after prolonged global cerebral ischemia in cats. Crit Care Med 16:964-971

Hossmann K (1998) Experimental models for the investigation of brain ischemia. Cardiovasc Res 39:106-120

Hossmann KA, Grosse-Ophoff B (1986) Recovery of monkey brain after prolonged ischemia. I. Electrophysiology and brain electrolytes. J Cereb Blood F Met 6:15-21

Hossmann KA, Zimmermann V (1974) Resuscitation of the monkey brain after 1h complete ischemia. I. Physiological and morphological observations. Brain Res 8:59-74

Hossmann K et al (1970) Recovery of neuronal function after prolonged cerebral ischemia. Science 168:375-376

Hossmann K et al (1973) Return of neuronal functions after prolonged cardiac arrest. Brain Res 60:423-438

Hossmann K et al (1975) Resuscitation of the monkey brain after one hour's complete ischemia. Brain Res 85:1-11

Jones PA et al (2004) Apoptosis is not an invariable component of in vitro models of cortical cerebral ischemia. Cell Res 14:241-250

Josselyn SA et al (2015) Finding the engram. Nat Rev Neurosci 16:52134

Kalimo H et al (1977) The ultrastructure of brain death. II electron microscopy of feline cortex after complete ischemia. Virchows Arch B Cell Pathol 25:207-220

Keidel WD (ed) (1975) Kurzgefasstes Lehrbuch der Physiologie. Thieme, Stuttgart

Klassen H et al (2004) Isolation of retinal progenitor cells from post-mortem human tissue and comparison with autologous brain progenitors. J Neurosci Res 77:334-343

Konishi Y et al (2002) Isolation of living neurons from human elderly brains using the immunomagnetic sorting DNA-linker system. Amer J Path 161:1567-1576

Latif M et al (2012) Skeletal muscle stem cells adopt a dormant cell state post mortem and retain regenerative capacity. Nat Commun 3:903

Lee K et al (2017) Derivation of Leptomeninges Explant Cultures from Post-mortem Human Brain Donors. J Vis Exp (119):55045. doi: 10.3791/55045

Leonard BW et al (1991) The influence of postmortem delay on evoked hippocampal field potentials in the in vitro slice preparation. Exper Neurol 113:373-377

Lipton P (1999) Ischemic cell death in brain neurons. Physiol Rev 79:1431-1568

Lisman J (2015) The challenge of understanding the brain: where we stand 2015. Neuron 86:864-882

Llorens-Bobadilla E et al (2015) Single-Cell Transcriptomics Reveals a Population of Dormant Neural Stem Cells that Become Activated upon Brain Injury. Cell Stem Cell 17:329-340

Lourbopoulos A et al (2015) Microglia in action: how aging and injury can change the brain's guardians. Front Cell Neurosci 9:54

Love S et al (2000) Neuronal death in brain infarcts in man. Neuropathol Appl Neurobiol 26:55-66

Lovell MA et al (2006) Isolation of neural precursor cells from Alzheimer's disease and aged control postmortem brain. Neurobiol Aging 27:909-917

Mansilla E et al (2013) Salvage of cadaver stem cells (CSCs) as a routine procedure: history or future for regenerative medicine. J Transplant Technol Res 3:118

Menzebach A et al (2010) A comprehensive study of survival, tissue damage, and neurological dysfunction in a murine model of cardiopulmonary resuscitation after potassium-induced cardiac arrest. Shock 33:189-196

Mayer UM, Sames K (1981) Oxygen consumption and aging in longterm cultures of lens epithelial cells. Albrecht v. Graefes Arch. Klin. Ophthalmol. 217: 117-124

Opelz G (1998) Cadaver kidney graft outcome in relation to ischemia time and HLA match. Transplant Proc 30:4294-4296

Ortega-de San Luis C, Ryan TJ (2922) Understanding the physical basis of memory: Molecular mechanisms of the engram. J Biol Chem 298:article 101866

Palmer TD et al (2001) Cell culture: progenitor cells from human brain after death. Nature 411:42-43

Pearson J et al (1977) Brain death: II. neuropathological correlation with the radioisotopic bolus technique for evaluation of critical deficit of cerebral blood flow. Ann Neurol 2:206-210

Pichugin Y (2006) Cryopreservation of rat hippocampal slices by vitrification. Cryobiology 52:228-240

Pichugin Y (2006a) Problems of long-term cold storage of patients' brains for shipping to CI. The Immotalist 38:14-20

Plug BC et al (2024) Human post-mortem organotypic brain slice cultures: a tool to study pathomechanisms and test therapies. Acta Neuropathol Commun 12(1):83. doi:10.1186/s40478-024-01784-1

Poo M-M et al (2016) What is memory? The present state of the engram. BMC Biology 14, art Nr 40

Pulsinelli WA (1982) Temporal profile of neuronal damage. Ann Neurol 11:491-498

Qi XR et al (2018) Alterations in the steroid biosynthetic pathways in the human prefrontal cortex in mood disorders: a post-mortem study. Brain Pathol 28:536–547

Radovsky A et al (1995) Regional prevalence and distribution of ischemic neurons in dog brains 96 hours after cardiac arrest of 0 to 20 minutes. Stroke 26:2127-2133

Ratych RE et al (1987) The primary localization of free radical generation after anoxia/reoxygenation in isolated endothelial cells. Surgery 102:122-131

Richter JD, Klann E (2009) Making synaptic plasticity and memory last: mechanisms of translational regulation. Genes Dev 23:1-11

Rupalla K et al (1998) Time course of microglia activation and apoptosis in various brain regions after permanent focal cerebral ischemia in mice. Acta Neuropathol 9:172-178

Safar P (1993) Cerebral resuscitation after cardiac arrest: research initiatives and future directions. Ann Emerg Med 22:324-349

Safar P et al (1976) Amelioration of brain damage after 12 minutes cardiac arrest in dogs. Arch Neurol 33:91-95

Saito T et al (2020) Isolation and culture of human adipose-derived mesenchymal stromal/stem cells harvested from postmortem adipose tissues. J Forensic Leg Med 69:101875. doi: 10.1016/j.jflm.2019.101875

Sames K (1980) Morphologische und histochemische Untersuchungen über das in-vitro und in-vivo Altern von Corneaendothel und Trabeculum corneosclerale. Universität Erlangen, Habilitationsschrift

Sames KH (2018b) Definitions of death. In Sames KH (ed) Applied human cryobiology, vol. 2, ibidem, Stuttgart

Sanal N (2004) Unique astrocyte ribbon in adult human brain contains neural stem cells but lacks chain migration. Nature 427:740-744

Schwartz PH et al (2003) Isolation and characterization of neural progenitor cells from post-mortem human cortex. J Neurosci Res 74:838-851

Senn P et al (2007) Robust postmortem survival of murine vestibular and cochlear stem cells. J Assoc Res Otolaryngol (JARO) 8:194-204

Shaffner DH et al (1999) Effect of arrest time and cerebral perfusion pressure during cardiopulmonary resuscitation on cerebral blood flow, metabolism, adenosine triphosphate recovery, and pH in dogs. Crit Care Med 27:1335-1342

Sheleg SV et al (2008) Stability and autolysis of cortical neurons in postmortem adult rat brains. Int J Clin Exp Pathol 1:291-299

Slivka A et al (1995) Cerebral edema after temporary and permanent middle cerebral artery occlusion in the rat. Stroke 26:1065-1066

Solenski NJ et al (2002) Ultrastructural changes of neuronal mitochondria after transient and permanent cerebral ischemia. Stroke 33:816-824

Sorrells SF et al (2018) Human hippocampal neurogenesis drops sharply in children to undetectable levels in adults. Nature 555:377-381

Takahashi M, Macdonald RL (2004) Vascular aspects of neuroprotection. Neurol Res 26:862-869

Verwer RWH et al (2002a) Tissue cultures from adult human postmortem subcortical brain areas. J Cell Mol Med 6:429-432

Verwer RWH et al (2002b) Cells in human postmortem brain tissue slices remain alive for several weeks in culture. FASEB J 16:54-60

Verwer RWH et al (2003) Post mortem brain tissue cultures from elderly control subjects and patients with a neurodegenerative disease. Exper Gerontol 38:167-172

Viel JJ et al (2001) Temperature and time interval for culture of postmortem neurons from adult rat cortex. Restoration of brain circulation and cellular functions hours post-mortem. Nature 568:336-343

Vrselia Z et al (2019) Restoration of brain circulation and cellular functions hours post-mortem. Nature 568:336-343

White RJ et al (1966) Prolonged whole brain refrigeration with electrical and metabolic recovery. Nature 209:1320-1322

Wolfs TG et al (2005) Apoptotic cell death is initiated during normothermic ischemia in human kidneys. Am J Transplant 5:68-75

Wu L et al (2008) Neural stem cells improve neuronal survival in cultured postmortem brain tissue from aged and Alzheimer patients. J Cell Mol Med 12:1611-1621

Wu S et al (1998) Reactive oxygen species in reoxygenation injury of rat brain capillary endothelial cells. Neurosurgery 4:577-583, discussion 584

Xu Y et al (2003) Isolation of neural stem cells from the forebrain of deceased early postnatal and adult rats with protracted post-mortem intervals. J Neurosci Res 74:533-540

Yakovlev AG, Faden AI (2004) Mechanisms of neural cell death: implications for development of neuroprotective treatment strategies. NeuroRx 1:5-16

Yamashima T, Oikawa S (2009) The role of lysosomal rupture in neuronal death. Prog Neurobiol 89:343-358

14 Cryonics can start to play along – intervention options after total organ failure

Summary

Medicine serves to Maintain life and health. Doctors try everything to preserve the life of a patient as long as there is a prospect of doing so and means of intervention are available. Cryonics is in the same sense an intervention in the dying process after organ failure, which serves the preservation of the still viable parts of living structures. It is supported by research into the processes of oxygen deprivation and organ failure as we have described above. What possibilities of intervention become visible and how can we develop them? On the one hand, how can we prevent harmful changes caused by oxygen deficiency and organ failure, and at the same time keep the harmfulness of our actions low? Effective measures are the throttling of chemical processes (e.g. by cooling), as well as the direct intervention in metabolic processes during cell death and harmful metabolic reactions e.g. by blocking metabolic pathways even after organ failure. It is important to allow continuous circulation (e.g. by dilution or removal of blood or by anticoagulation).

14.1 Cooling still the best conservation method even after "death".

We have already discussed above the beneficial effects of cooling and how it can be used as a means of resuscitation in emergency medicine, while characterizing it as the first step toward cryonics.

When the blood circulation stops, the metabolism, which produces heat, decreases and heat is no longer transported through the vessels. The body cools down to the ambient temperature. Unfortunately, this occurs very slowly at the beginning, so that one must help as early as possible (Best 2008; see also Best Physical parameters of cooling in cryonics https://www.benbest.com/cryonics/cooling.html).

A small decrease in brain temperature has a beneficial effect, even after the energy of the cells is depleted. Cooling produces a great effect in improving brain preservation.

Rats that were perfused again at 36°C after 15 minutes of oxygen deprivation had three times as many of the very harmful hydroxyl radicals one hour later than rats that were not perfused again. But rats that were perfused at 30°C had only half as many hydroxyl radicals (Ki et al. 1996).

Ben Best estimates that the damage of storing brain tissue for five hours at room temperature (with the body cooling to room temperature) is equivalent to about 5 days in a hospital cold room at 2-7°C. Even better would be storage at water/ice temperature (review Best B: Quantifying ischemic damage for cryonics rescue. https://www.benbest.com/cryonics/IR_Damage.html).

Despite the positive effect against the damage of circulatory arrest, 4°C for more than one to two days, for example,—as it was mentioned above—leads to lower survival of kidneys in organ protection solution than with shorter periods of cooling (Opelz 1998). Cell membranes may become permeable as a result, allowing an influx of ions and leading to osmotic changes with water influx. The interior of the cell may become acidic, proteins may be altered and cell suicide may be initiated, reactive oxygen compounds may be increased, energy reserves may dwindle, and the cytoskeleton of cells may be damaged (Rubinsky et al. 2003). However, such cold oxygen starvation is survived by some of the cells for days (Best 2012). The above-mentioned studies by Pichugin on brain slices and cultures of brain cells showed a slow change in the ratio of potassium to sodium ions at cool temperatures (4°C) (Pichugin 2006a)

According to a cryonic rule of thumb, 45-60 minutes of circulatory arrest in the warm without anticoagulant medication no longer allows circulation.

In the case of cooling, the time the patient has spent at room temperature before cooling and during transport plays an important role. The cooling then has a preserving effect but at reperfusion with blood flowing again, the inner wall cells of the blood vessels are significantly more damaged after cold circulation stop than after circulation stop at normal temperature.

After 10 hours at refrigerator temperature, the flow of vitrification solution is likely to be too difficult, especially since the solution is more viscous in the cold and probably more difficult to pass through the capillaries of the circuit. Bacterial growth must also be expected here (De Groot, Rauen 2007; see also Best B Quantifying ischemic damage for cryonics rescue. https://www.benbest.com/cryonics/IR_Damage.html<).

Thus, low temperatures above freezing are very favorable at the start of cooling, but longtime storage at such temperatures should be avoided.

14.2 Prevention of vascular occlusion

The blockage of blood flow after organ failure ("no reflow") can also be influenced. We address this at various points in this book.

Even after a short period of blood flow stop, it becomes more difficult to perfuse the circulation. Increasingly higher pressure can be applied. The simplest means of resuscitation, cardiac massage, is no longer effective after only 6 minutes because it generates too little pressure.

Resistance in the vessels increases as early as 6 minutes after blood flow stops.

> Background information
> Until then, cardiac massage (esp. by a mechanical device) can produce a blood pressure of 24mm Hg. After 6 minutes, the blood pressure in the brain must be increased to 35mmHg to achieve perfusion and after 12 minutes, even this pressure no longer allows perfusion (Shaffner 1999). After 30 minutes of hemostasis, cats need 100mm Hg for flow to occur (Iijima 1993).

Since blood cells contribute to the damage upon reperfusion, their removal by a washout prior to cold blood flow arrest (transport in ice) may be beneficial. It can reduce the damage caused by reperfusion in a district of the rat brain with vascular occlusion (Ding et al. 2002). Red blood cells, as mentioned, cluster together when blood is stagnant, creating an obstruction (Leonov et al. 1992). Dilution of the blood (hemodilution) is helpful here. Dextran, for example, can be used for this purpose, which increases the osmotic pressure in the vessels and thus draws water into them. It thus prevents swelling of the tissue.

The prevention of blood clots is important. We have already discussed anticoagulation in general terms.

A less hazardous inhibition of clotting is done, for example, by aspirin and vitamin E.

> Background information
> The anticoagulants also include proteoglycan products such as the glycosaminoglycan of cartilage chondroitin sulfate. It is closely related to heparin, which is also a proteoglycan product. In heart infarction, glycosaminoglycans can prevent recurrences and they even show an effect against Alzheimer's disease (see in Ban, Lehmann 1989). However, they are not frequently used for this purpose.

These substances in contrast to heparin are to be considered like food supplements and a terminally ill patient can take them himself if he does not already have a coagulation disorder.

Heparin (Hep) can only prevent blood clots. It cannot dissolve clots that have formed. It is used by Cryonics Institute (CI) to inhibit blood clotting.

> Background information
> It is also anti-inflammatory (including modified forms e.g. O-sulfated heparin) and is an anti-histamine agent (Stullken, Sokoll 1976; Wang et al. 2002). Hep can be given up to 7 minutes after cardiac arrest, according to cryonics executive. However, no effect was found in an experiment in which blood flow was stopped for 15 minutes at 15 minutes after foregoing heparin administration in rabbits. (Fischer, Ames 1972).

Streptokinase can dissolve clots. However, the effect decreases with longer existing clots. It also has a preventive effect (Verstruete 1985) but can lead to severe bleeding if not controlled. Streptokinase can dissolve clots up to about 15 min after their formation.

Plasmin is a protein-digesting enzyme that attacks the fibrin fibers in blood clots. It is initiated by tissue plasminogen activator (tPA). Plasminogen activator has similar capabilities to streptokinase. Unfortunately, it can increase the permeability of the capillaries in the brain, allowing water to enter the tissues (brain swelling). It is also very expensive (Yepes et al. 2009). Treatment with tPA (see above) also decreases in effect with the duration of blood flow arrest (Hacke 2004). Plasmin unfortunately also attacks vessel walls and the blood-brain barrier (Clark et al. 2000; Pfefferkorn, Rosenberg 2003), which is particularly critical during cryonic perfusion and late treatment.

If such drugs are given during life as a preventive measure, which is also possible for cryonic candidates, the physician must decide whether a coagulation disorder is already present that could be dangerously aggravated.

Survival had been increased in experiments with methods in which heparin plays an important role.

> Background information
> Heparin was also used in the prolongation of survival of dogs discussed above (Safar 1993; Safar et al. 1976; Shaffner 1999). With hypothermia, elevation of blood pressure, and blood thinning, improvement was still possible, but not complete recovery of the brain (Safar et al. 1996). Mike Darwin used similar methods to increase survival time in circulatory arrest to 17 minutes. Abundant antioxidants were administered but without blood pressure elevation. The dogs were kept at (human) body temperature, vasopressin was beneficial (see also Best B Quantifying ischemic damage for cryonics rescue http://www.benbest.com/cryonics/IR_Damage.html). However, in more recent experiments, the benefit of this approach was not clear (de Wolf, De Wolf 2013).
> The administration of vitamin E injected into the veins 30 min before blood flow stops reduces damage to nerve cells as a result of a blood flow stop (Yamamoto et al. 1983). Vitamin E also inhibits clotting, and unlike aspirin, it does not cause petechial stomach bleeding. Many fish oils have the same effect besides influencing atherosclerosis via blood lipids (Charnock et al. 1992).
> A cryonics candidate can already take vitamin E as a dietary supplement as a preventive measure to prevent clotting after death.

14.3 Agents against cell death

Many cells in our body are easily replaceable, but whether this works for the intricately networked brain cells is unclear. Therefore, ways to protect the cells are highly interesting.

Possibilities to influence the suicide of brain cells already exist. However, the complete stop of suicide and a targeted recovery of cells in which suicide has already progressed are still a major problem.

It may be an advantage here that suicide progresses slowly in case of cardiac arrest. This could be — as mentioned — because energy is missing.

One way to promote the survival of neurons in culture was to keep the cells together with rat embryonic stem cells (cocultivation). This involved placing the two types of cells together in special vessels with a separating semipermeable membrane. Active substances from the stem cells migrate through the membrane and can be taken up by the nerve cells. In this experiment, the human cells were taken from human brain cortical tissue up to 9.5 hours after death. During this process, the brains spent 1-2 hours at room temperature during transport. The patients were up to 94 years old.

> Background information
> The number of dead cells averaged about 17 out of 26 (per cmm). Slightly more than 3 of the survivors were healthy, the others damaged. In contrast, in the cells that were held together with embryonic rat cells, about 6 of 16 of the neurons were dead. Of the others, about 6 were healthy and about 4 were damaged. That is, the number of dead cells decreased from about 65% to 38% by growing together with the embryonic cells, and the number of damaged cells also decreased (Wu et al. 2008).

Thus, in this experiment, not only were living brain cells obtained from deceased persons, but a means was also found to increase the number of surviving cells. However, this means that the additional survivors were still alive when they were extracted. If the substances of the rat cells that promote survival are analyzed, they possibly can also be used after cardiac arrest to keep more cells alive. So, there are completely new possibilities developing here.

Other methods try to intervene in the suicide of the cells. They target the enzymes, for example. Inhibition of the already mentioned enzyme caspase interfering with cell suicide protected certain groups of neurons but not others in an experiment with artificially induced stroke (Zhan et al. 2001). In another experiment, an inhibitor of the aforementioned dangerous caspase-3, named Z-DEVD-FMK, promoted the survival of neurons in that important part of the brain, the hippocampus, in rats during arrest of blood flow (Chen et al. 1998).

> Background information
> Cell suicide due to blood flow arrest/reperfusion can also be influenced by Propofol (Javadov et al. 2000; Polster et al. 2003).

Free radicals play a negative role in different types of damage. To intervene here, the free radical scavenger NXY-059 was administered to rats 15 minutes after a 2-hour blood arrest. The blood flow arrest with infarction affected a focal district. NXY-059 decreased the extent of this focal area (Sydserff et al. 2002).

> Background information
> Coenzyme Q10 acts against oxygen products and damage by circulatory arrest. It protects the inner vessel wall cells during reperfusion (Yokoyama et al. 1996). Of patients with cardiac arrest who received Q10 within 6 hours 68% survived, while survival of untreated ones was only 30% (Damian et al. 2004). Curcumin has a strong effect against free radicals. It can scavenge peroxinitrite and inhibit the enzyme nitrite oxidase (Lim et al. 2001). It is contained in a commercial spice (curcuma, turmeric). It is also reported to be effective against malignant tumors (Rao et al. 1999) and usable as a teeth whitener. It can be taken in large doses with food. It is also present in curry in high percentage. By the way, curry and tomato (also protective against tumors) are the healthy parts of curry sausage (especially without sausage).

Animal studies have demonstrated beneficial effects of the widely praised antioxidants—which render dangerous oxygen products harmless, as well as vitamin E (Hara et al. 1990), melatonin, deprenyl, and PBN (El-Abhar et al. 2002; Folbergrova et al. 1995; Kuhmonen et al. 2000; Wang et al. 2002).

Hydroxyl radicals, which are produced when blood flow stops, can be scavenged by melatonin. Melatonin acts by releasing electrons. Its effect is three orders of magnitude greater than that of Vit E (Chyan space et al. 1999). Melatonin is also said to have an anti-aging effect. When this first became known, it sold out in a short period at a time when gerontology was neither taken sufficiently seriously nor promoted. Pretreatment of gerbils with melatonin 30 minutes prior to reflow after stop of blood circulation significantly reduced brain damage (Cuzzocrea et al. 2000). Similar results were obtained in rats (Pei et al. 2003).

Deprenyl also has a beneficial effect during stop of blood flow. It can protect cell cultures from cell suicide (Maruyama, Naio 1999). Pretreatment of gerbils with 0.25mg/Kg deprenyl two weeks before blood flow arrest reduced damage to cells in the sensitive brain part hippocampus (Kuhmonen et al. 2000). Deprenyl has life-prolonging effects in mice and Parkinson's patients as well (see Sames 1991).

Insulin promotes survival of vulnerable neurons in total oxygen starvation (Sanderson et al. 2009).

Locally acting anesthetics decrease brain damage by blocking sodium channels and indirectly by decreasing electrical activity and metabolic rate (Urenjak, Obrenovitch 1996).

> Background information
> It is possible to increase the function of the kidneys after cold blood circulation arrest and reperfusion. Substances that release carbon monoxide and favorably influence cellular respiration have a protective and vasodilatory effect (Sandouka et al. 2006).
> Lipoic acid acts against damage caused by reperfusion directly and by protecting glutathione and peroxinitrite as well as inhibiting the enzyme xanthine oxidase (see above) (Packer et al. 1995; Rezk et al. 2004).

We can learn a lot from the treatment of stroke, which results from occlusion of individual brain vessels.

Some of the above-mentioned damage due to oxygen deprivation has been studied in stroke or vascular occlusion in animal experiments (see above). Stroke and other vascular occlusions thus also become models for the study of the ultimate cessation of blood flow. Stroke therapy is about reducing the spread of damage and counteracting vascular occlusion.

In stroke, however, the tissue remains at body temperature and it is surrounded by healthy tissue. Reactions such as inflammation (see e.g. Lourbopoulos et al. 2015) are possible, which are only found for a few minutes in the case of total blood flow arrest (e.g. heart failure) due to lack of energy. In any case, stroke can teach us some things that happen in these first minutes.

The action of drugs can be tested in stroke, but having said that, it is clear that the results cannot be simply transferred to the final arrest of blood flow, as it is seen in cryonics.

In stroke, the most effective anticoagulants must be used quickly. The aforementioned tissue plasmin activator (tPA) can dissolve clots in the first 3 hours,—but especially in the first 15 minutes—after blood flow stops. For the prevention and dissolution of clots, a patient with stroke must therefore be treated as soon as possible preferably (for infusions and monitoring) in a clinic. tPA can, for example—given until a few hours after a stroke—guarantee chances of a good outcome for 3 months (Gumbinger CH 2014).

Special blockers can, for example, prevent harmful active substances from binding to the cell. This inhibits substances such as the so-called tumor necrosis factor and other agents causing damage. Certain blockers also act against cell suicide in stroke (Martin-Villaba et al. 2001; Tuttolomondo et al. 2009).

One would think that drugs that inhibit calcium influx in cells would be beneficial in overexcitation

> Background information
> Inhibition is conceivable, for example, by blocking the NMDA binding site. But the respective agents fail in clinical trials. In animal experiments, one then sees that an effect lasts only in the first 4 minutes. Blockers of another access, the L channel do not change this. The T channels remain open and calcium flows into the cell anyway because the calcium pumps fail. The influx of calcium into the cell after circulatory arrest can be blocked by nimodipine (Babu et al. 2011). In dogs, pretreatment with nimodipine worked. They recovered from oxygen deprivation at a rate of 80%. In contrast, 86%

died without treatment (Ginsberg 1988). Drugs that block the L channel for calcium are the di-hydro-pyridine (DHP) derivatives, but their beneficial effect is more likely due to a side effect. Namely, they also dilate the small arterial branches (arterioles). Failure of the potassium/sodium pump due to lack of energy leads to high activity of another system for balancing the molecules, the sodium/potassium exchanger. This, however, consumes further energy from the chemical store (see also Best B Ischemia and reperfusion injury in cryonics. https://www.benbest.com/cryonics/ischemia.html).

Overall, a lot of research is still needed here.

A completely different possibility is the repair after a stroke with loss of nerve cells. Here, embryonic stem cells were used, which can develop into nerve cells (Bühnemann et al. 2006). To what extent and in what way a brain can be repaired by stem cells in the case of absolute oxygen starvation is an interesting question. There is no shortage of repair cells.

Literature

Babu CS, Ramanathan (2011) Post-ischemic administration of nimodipine following focal cerebral ischemic-reperfusion injury in rats alleviated excitotoxicity, neurobehavioral alterations and partially the bioenergetics. Int J Dev Neurosci 29:93-105

Ban TA, Lehmann HE (eds) (1989) Diagnosis and treatment of old age dementia. Karger, Basel

Best BP (2008) Scientific justification of cryonics practice. Rejuvenation Res 11:493-503

Best BP (2012) Vascular and neuronal ischemic damage in cryonics patients. Rejuvenation Res 15:165-169

Bliss LA et al (2017) Use of Postmortem Human Dura Mater and Scalp for Deriving Human Fibroblast Cultures. PLOS ON September 27, https://doi. or10.1371/journal.pone.0045282

Bühnemann C et al (2006) Neuronal differentiation of transplanted embryonic stem cell-derived precursors in stroke lesions of adult rats. Brain 129:3238-3248

Charnock JS et al (1992) Dietary modulation of lipid metabolism and mechanical performance of the heart. Mol Cell Biochem 116:19-25

Chen J et al (1998) Induction of caspase-3-like protease may mediate delayed neuronal death in the hippocampus after transient cerebral ischemia. J Neurosci 18:4914-4928

Chyan et al (1999) Potent neuroprotective properties against the Alzheimer b-amyloid by an endogenous melatonin-related indole structure, indole-3-propionic acid. J Biol Chem 274:21937-21942

Clark WM et al (2000) The rtPA (alteplase) 0- to 6-hour acute stroke trial, part A (A0276g) results of a double-blind, placebo-controlled, multicenter study. Stroke 31:811-816

Cuzzocrea S et al (2000) Protective effects of melatonin in ischemic brain injury. J Pineal Res 29:217-227

Damian MS et al (2004) Coenzyme Q10 combined with mild hypothermia after cardiac arrest. Circulation 110:3011-3016

De Groot H, Rauen U (2007) Ischemia-reperfusion injury: processes in pathogenetic networks: a review. Transplant Proc 39:481-484

De Wolf A, De Wolf C (2013) Human cryopreservation research at advanced neural biosciences. In Sames KH (ed) Applied Human Cryobiology, vol. 1, ibidem, Stuttgart, pp 45-59

Ding Y et al (2002) Pre-reperfusion flushing of ischemic territory: A therapeutic study on ischemia-reperfusion injury in stroked rats using histological and behavioral assessments. J Neurosurg 96:310-319

El-Abhar HS et al (2002) Effect of melatonin and nifedipine on some antioxidant enzymes and different energy fuels in the blood and brain of global ischemic rats. J Pineal Res 33:87-94

Fischer EG, Ames A (1972) Studies on mechanisms of impairment of cerebral circulation following ischemia: effect of hemodilution and perfusion pressure. Stroke 3:538-542

Folbergrova J et al (1995) N-tert-butyl-a-phenylnitrone improves recovery of brain energy state in rats following transient focal ischemia. Proc Natl Acad Sci USA 92:5057-5061

Ginsberg MD (1988) Efficiency of calcium channel blockers in brain ischemia—a critical assessment. In Krieglstein J (ed) Pharmacology of cerebral ischemia. Proceedings of the second international symposium on pharmacology of cerebral ischemia. Wissenschaftliche Verlagsgesellschaft, Stuttgart pp 65-67

Gumbinger CH et al (2014) Time to treatment with recombinant tissue plasminogen activator and outcome of stroke in clinical practice: retrospective analysis of hospital quality assurance data with comparison with results from randomised clinical trials. BMJ 2014;348:g3429, doi: 10.1136/bmj.g3429

Hacke W et al (2004) Association of outcome with early stroke treatment: pooled analysis of ATLANTIS, ECASS, and NINDS rt-PA stroke trials. Lancet 363:768-774

Hara H et al (1990) Protective effect of alpha-tocopherol on ischemic neuronal damage in the gerbil hippocampus. Brain Res 510:335-338

Iijima T et al (1993) Brain resuscitation by extracorporeal circulation after prolonged cardiac arrest in cats. Intensive Care Med 19:82-88

Javadov SA et al (2000) Protection of hearts from reperfusion injury by propofol is associated with inhibition of the mitochondrial permeability transition. Cardiovas Res c45:360-369

Kuhmonen J et al (2000) The neuroprotective effects of (-)deprenyl in the gerbil hippocampus following transient global ischemia. J Neural Trans 107:779-786

Leonov Y et al (1992) Hypertension with hemodilution prevents multifocal cerebral hypoperfusion after cardiac arrest in dogs. Stroke 23:45-53

Lim GP et al (2001) The curry spice curcumin reduces oxidative damage and amyloid pathology in an Alzheimer transgenic mouse. J Neurosci 21:8370-8377

Lourbopoulos A et al (2015) Microglia in action: how aging and injury can change the brain's guardians. Front Cell Neurosci 9:54

Martin-Villalba A et al (2001) Therapeutic neutralization of CD95-ligand and TNF attenuates brain damage in stroke. Cell Death Differ 8:679-686

Maruyama W, Naoi M (1999) Neuroprotection by (-)-deprenyl and related compounds. Mech Ageing Dev 111:189-200

Opelz G (1998) Cadaver kidney graft outcome in relation to ischemia time and HLA match. Transplant Proc 30:4294-4296

Packer L et al (1995) Alphal-lipoic-acid as a biological antioxidant. Free Radic Biol Med 19:227-250

Pei Z et al (2003) Melatonin reduces nitric oxide level during ischemia but not blood-brain barrier breakdown during reperfusion in a rat middle cerebral artery occlusion stroke model. J Pineal Res 34:110-118

Pfefferkorn T, Rosenberg R (2003) Closure of the blood-brain barrier by matrix metalloproteinase inhibition reduces rtPA-mediated mortality in cerebral ischemia with delayed reperfusion. Stroke 34:2025-2030

Polster BM et al (2003) Inhibition of Bax-induced cytochrome-c release from neural cell and brain mitochondria by dibucaine and propranolol. J Neurosci 23:2735-2743

Rao ChV et al (1999) Chemoprevention of colonic aberrant crypt foci by an inducible nitric oxide synthase-selective inhibitor. Carcinogenesis 20:641-644

Rezk BM et al (2004) Lipoic acid efficiently protects only against a specific form of peroxynitrite-induced damage. J Biol Chem 279:9693-9697

Rubinsky B (2003) Principles of low temperature cell preservation. Heart Fail Rev 8:277-284

Safar P (1993) Cerebral resuscitation after cardiac arrest: research initiatives and future directions. Ann Emerg Med 22:324-349

Safar P et al (1976) Amelioration of brain damage after 12 minutes cardiac arrest in dogs. Arch Neurol 33:91-95

Safar P et al (1996) Improved cerebral resuscitation from cardiac arrest in dogs with ibld hypothermia plus blood flow promotion. Stroke 27:105-113

Sames K (1991) Molekularbiologische Aspekte des Alterns. In Expertisen zum 1. Teilbericht der Sachverständigenkomission zur Erstellung des 1. Altenberichts der Bundesregierung. DZA, Berlin

Sanderson TH et al (2009) Insulin activates the PI3K-Akt survival pathway in vulnerable neurons following global brain ischemia. Neurolog Res 31:947-958

Sandouka A et al (2006) Treatment with CO-RMs during cold storage improves renal function at reperfusion. Kidney Int 69:239-247

Shaffner DH et al (1999) Effect of arrest time and cerebral perfusion pressure during cardiopulmonary resuscitation on cerebral blood flow, metabolism, adenosine triphosphate recovery, and pH in dogs. Crit Care Med 27:1335-1342

Stullken EH Jr, Sokoll MD (1976) The effects of heparin on recovery from ischemic brain injuries in cats. Anesth Analg 55:683-687

Sydserff SG et al (2002) Effect of NXY-059 on infarct volume after transient or permanent middle cerebral artery occlusion in the rat; studies on dose, plasma concentration and therapeutic time window. Br J Pharmacol 135:103-112

Tuttolomondo A et al (2009) Neuron protection as a therapeutic target in acute ischemic stroke. Curr Top Med Chem 9:1317-1334

Urenjak J, Obrenovitch TP (1996) Pharmacological modulation of voltage-gated Na+ channels: a rational and effective strategy against ischemic brain damage. Pharmacol Rev 48:21-67

Verstruete M (1985) As little as 250,000 IU of streptokinase could reduce plasma fibrinogen to 30% of the starting level. Eur Heart J 6:586-593

Wang L et al (2002) Heparin's anti-inflammatory effects require glucosamine 6-O-sulfation and are mediated by blockade of L- and P-selectins. J Clin Invest 110:127-136

Wu L et al (2008) Neural stem cells improve neuronal survival in cultured postmortem brain tissue from aged and Alzheimer patients. J Cell Mol Med 12:1611-1621

Yamamoto M et al (1983) A possible role of lipid peroxidation in cellular damages caused by cerebral ischemia and the protective effect of alpha-tocopherol administration. Stroke 14:977-982

Yokoyama H et al (1996) Coenzyme Q10 protects coronary endothelial function from ischemia reperfusion injury via an antioxidant effect. Surgery 120:189-196

Zhan et al (2001) Both caspase-dependent and caspase-independent pathways may be involved in hippocampal CA1 neuronal death because of loss of cytochrome c from mitochondria in a rat forebrain ischemia model. J Cerebral Blood Flow Metabolism 21:529-540

15 Restoration

Summary

As far as human application is concerned, the discussed possibilities of rewarming and restoration including nanotechnology are still futurology. Possible approaches are presented here. However, in some tiny creatures, resuscitation works, as we have seen above. In animal experiments and cryopreservation of small organs there are successful steps also in the laboratory.

15.1 Rewarming, recovery and reanimation

The brain even after organ failure still has means of repair and rejuvenation. But here the blood circulation is missing.

Knowledge of the so-called connectome (this roughly refers to the network of all cells in the brain (McIntyre, Fahy 2017; Mikula 2016) may help in a definitive reconstruction of the organ in the future.

As mentioned, glia cells the astrocytes can also replace brain neurons. Glial cells are a type of special cells that play the role of connective tissue cells in the nervous system. Astrocytes are considered as a kind of nurses of the brain neurons, which regulate e.g. the supply of substances to the neurons (Kriegstein, Alvarez-Buylla 2009; Merkle et al. 2004).

It is astonishing that neurons formed from glia cells, which are brought into brain centers, with their extensions grow exactly to the same subordinate brain centers as the cells they replaced almost as if they could read thoughts (Ideguchi 2010). They thus perform reconstruction work on their own.

Other stem cells, e.g., from bone marrow, can also transform into neurons (Mezey et al. 2003).

We discussed stem cell survival in organ failure above.

We can expect that an anatomically restored brain will also function normally again.

The brain survives even major damage. It is known from pathology, for example, that very large hemorrhages, if the brain survives them, occasionally do not cause any symptoms and are only discovered during a dissection. Memory content that is stored multiple times may also play a role in recovery. In this way, the information is retained even if parts of brain tissue are damaged. In any case, the adaptability of brain cells (plasticity) plays a role (Gertz 1989). Cells can functionally substitute and replace each other. This ability of the brain raises further hope for possible recovery (Carmichael 2006; Chen et al. 1998; delete Dancause et al. 2005; Nudo 2007; Wu et al. 2008). A glimmer of hope has been raised by experiments on the nematode Caenorhabditis elegans. The animals can be cooled to cryogenic temperatures and reawakened, as mentioned. Animals were trained to recognize a chemical signal and this ability (a type of memory response) survived cryopreservation (Vita-More et al. 2015).

15.2 Thawing and reanimation

The difficulties of thawing human bodies have already been discussed. Reparative and regenerative medicine may be able to contribute to recovery (Sames 2000a).

We have mentioned elsewhere (Sames 2000; 2013; 2013a) that eliminating the changes caused by aging and disease that medicine must capitulate to when a person passes away may require the work of generations. Such far-reaching projects as healing to complete recovery without permanent consequences (such as scars), stopping aging, or even rejuvenation cannot be achieved in a few decades (if ever). However, we do not see a fundamental obstacle here either. The proof of impossibility has often been claimed, but not provided.

It is therefore permissible to start developing the necessary procedures and planning projects in this direction. In fact, many are already underway.

In resuscitation, we hope that it will be possible to thaw and repair one organ at a time. Deep freezing of the heart only in the body of rats has already been practiced (Leunissen et al. 1968). In the experiments described above with cooling small rodents to temperatures of 0°C or slightly lower, the heart region was preferentially warmed during resuscitation without harming the animals. One could perhaps stabilize the organs during thawing and insulate them as well as possible from the rest of the body. For thawing and repair of the whole body at the same time, one would possibly need a very large team and a correspondingly complicated technique, or even just a working rewarming. For rapid thawing of the body without additional damage, perhaps a method with very rapid heating can be developed (see above). One could then even try to repair an already revived body. Another possibility would be the repair without thawing process in the frozen state (see below).

It is encouraging that many cells are still alive during cryopreservation (even if this occurs after about 8 hours). However, it is questionable how many of these cells would still be alive after thawing if a whole human body was vitrified. But those that are conveniently located to the vessels and optimally vitrified could survive. Such cells could potentially be used as stem cells. Various stem cells are also themselves located in close proximity to vessels. Perhaps they could be left in the tissue and stimulated to repair. It is also to be expected that methods will be developed to create the body's own repair cells, which require nothing other than intact DNA. This can be found in abundance in the body of a cryonics patient and is probably sufficiently preserved.

It is favorable that the tissue image and the cell forms in the organs are preserved long after death. This is the basis for the diagnosis made by pathologists and forensic pathologists after death. Microscopic examinations are also performed. Sometimes even microscopic preparations are made for medical teaching from tissue samples of deceased people—even from nerve tissue. It seems strange when medical doctors, whom we have trained in anatomy on cadavers and microscopic preparations, where death occurred a long time ago, do not perceive this fact and adhere to adventurous views about death (at least this is how it occasionally gets into the public domain).

Maybe we should start teaching theory of survival to students. The first lesson would be about why you can still recognize a corpse.

It is known that tissue images after deep-freezing at very high magnification sometimes reveal severe damage, but it may be possible to reconstruct the original condition in a model. Probably this can be done without major problems. The model could be used to target stem cells in the damaged tissue for repair. The preserved body structures can serve as a raw model. However, rejuvenation would require repairing the body cell by cell or molecule by molecule. The best means we have today to do this, such as genetic engineering, stem cell technology and bio-printers, are still far from this goal. It is especially difficult to preserve and restore the individual body and the individual ego consciousness. One can — as said — expect that the restoration of the structures will also make the body function again if energy is provided.

A positive outlook may now be opened up by a repair using nanotechnology.

15.3 The finishing touch: Nano repair

As a student, Dr. Siegfried Stoll thought intensively about the rejuvenation of a body atom by atom, after we realized that the means available today do not offer a short-term method of prolonging life. He imagined a device he called an atom shifter as a hypothetical means of doing this. With this one would shift one atom of a tissue after the other spatially a little bit and thereby eliminate wrong atoms and wrong connections, so that on the other side the healthy body grows up. He was electrified when he came across the books of K. Eric Drexler. We spent a lot of time in the 1980s speculating about repairing humans with the help of nanotechnology. These thoughts and other approaches from the literature will be incorporated into the following section (Drexler 1981; Drexler, Peterson 1994; Mathwig 2018; Mathwig, Sames 2013; Merkle 1992).

Nanotechnology perfected entirely in the biological sense could solve our reanimation problems and perhaps perform repair while still in cryostasis.

According to Ben Best, nanotechnological recovery from the glassy state can be represented as follows (Best 2013b).

Nanorobots (see above) first remove the water and other contents from the blood vessels. All body tissues can then be reached via the freed blood vessel. Some kind of nano-rails could clear out the contents of the blood vessels in a similar way to assembly lines.

A new alternative is provided by perfusion of organ vessels with solutions delivering nano particles. The particles remaining in the vessels—including capillaries—can be electromagnetically warmed resulting in resolution of ice or vitrified fluid. Thus, organ circulation is restored (s. 9.3.3).

By nano tech, if blood vessels have a leak or are blocked, they can be connected by artificial tunnels. That way, blood can continue to flow. Artifacts should not be confused with original structures. A nano-arm as described by Eric Drexler in his book: Nanosystems (1992) would need 4 million atoms and be 100nm long and 50 nm wide. It could move a lot of different tools to manipulate molecules to bring them directly into chemical bonds.

What chemical reactions achieve by chance in solutions with masses of molecules, nanotechnology could do specifically with individual molecules. We must expect physics to challenge biochemistry for the lead in this field. Today, chemical reactions are due to random hits of molecules that depend on concentration. Reactions are thus brought about in a relatively slow and untargeted manner. Nanotechnology could perform calculable targeted manipulations molecule by molecule (Drexler, Peterson 1994). Tools for "digging" need be of little complexity. Nanorobots would have to be much larger with propulsion, power, and information-processing parts.

A complete nanorobot (assembler) could be a few thousand (5000-10000) nanometers (nm) or 5-10 micrometers (um) in size. This is, of course, only an estimate at the near-real time level. A capillary has a diameter of about 7um i.e. 7000 nm. Ralph Merkle estimated that 3200 million "nanorobots" weighing a total of 53g could repair a cryonics patient in about 3 years (Merkle 1994; 1994a).

Merkle and Freitas (2008) proposed that the nanorobots be equipped with electrostatic motors. Generators and rotors should be electric rather than magnetic (tiny charged moving platelets are easier to make than tangles and small iron cores). Magnetic properties go badly with decrease in size (this is to say that magnetic motors of molecular dimensions would not work). Electrostatic properties then behave positively to reduction of size. Electrostatic transducers (actuators) are already in use in microelectromechanical systems (MEMS) (Fennimore et al. 2003).

High-density nanobatteries could provide energy for days, and recharging stations could be located throughout the patient. Alternatively, nanotube cables could conduct energy from outside to the patient. These could also serve to relay information data and could reunite fracture edges. Scanning and image processing could show what needs to be worked on. Replacement is often preferable to repair because it is easier to rebuild even whole organs, for example. Healing of diseases and rejuvenation would also be achieved. The brain, of course, would have to be repaired in its original personal form. But individual components could also be completely replaced in the brain, e. g. cell organelles and many large molecules. DNA could be restored and, if necessary, modified to intervene in healing and repair. However, subsequent processes (epigenetic processes) of DNA action could be crucial in restoring cell connections (synaptic structures). These cannot be read on the DNA. The connections (synapses) of nerve cell processes should not only be restored in their structure, but also their content of transmitter substances. The nature of these substances and their distribution in the cell and its environment must therefore be restored to the initial state. There are more than 40 transmitters. However, from the connections to other cells, it is possible to infer to some extent which transmitter substances are used, since different functions with special nerve connections use very specific transmitters.

Unlike in the living, operations would not lead to inflammation or defense reactions, nor would they disrupt any functions. One could therefore transplant and reconstruct several organs at the same time as in a construction kit. In this respect, organ failure and death would be the chance for rejuvenation.

Removal of ice crystallization nuclei, most of which are expected to be outside the cells, would be part of the repair.

The restored blood vessels would contain fresh cryoprotectant, water, plasma, and blood cells without the ice crystallization nuclei. No special repair above cryogenic temperatures would be necessary. The method would be performed on the frozen (and thus protected) body.

A large computer of the future, whose programs have the coordinates of all the molecules in a normal human body, could contribute to repair in a controlling way and program the nanorobots.

This would also give the possibility to keep the rejuvenated body youthful Using methods which resemble the procedure developed by the nature itself: a continuous molecular renewal based on a program. The natural renewal remains natural and can be supplemented and strengthened in such a way that aging no longer occurs and damage is eliminated. The chance to participate in such a technology in the future could be opened by cryonics. We certainly do not yet know all the hurdles and possible disasters we will stumble into. It is good to have this in mind.

One way to program the large-scale computer would be to split the brain in half and further halve it down to the smallest units, digitally recording the state after each split until the brain is fully digitally recorded (Merkle 1994; -1994a). Piece-by-piece cutting with a nano-microtome could serve digital capture (Mikula 2016; Mikula et al. 2015). Ideally, nanorobots could be used to "map" the organism rather than repair it by sending the conditions analyzed in the field to the large-scale computer. The complete digital data including those of spatial molecular arrangements could then be used for reconstruction. They could determine molecular complexes that deviate from the ideal body in the computer and preserve them individually or recognize them as defects and repair them.

A major problem could be the drive energy for billions of small robots. In any case, the heat generated must be taken into account. If one reduces the number of nanorobots, less energy but more time is needed.

The great challenge of nanotechnology today is still the difference in size between tool and object (nano jump). Molecules are of a size that allows their direct observation with today's means only in individual — favorably located — objects.

A co-inventor of the tunneling microscope in Switzerland compared the current state of affairs to trying to move a grain of sand in a targeted manner using the tip of the Matterhorn as a tool. E. Drexler's original concept is to use tools to build smaller tools, then use those tools to build even smaller tools, and so on down to the nanoscale (millionths of a millimeter).

Freitas has now updated the methods in great detail (Freitas 2022)

In order to maintain the youthful state of a body following the restoration, damage that threatens our self-confidence, peculiarities or functions would have to be constantly preventively repaired in the way indicated above in the life after thawing.

Overall, the aging changes would in fact be constantly ongoing in accordance with the organ differentiation hypothesis (Sames 2013) of aging. Nanobots would therefore have to migrate ever and anon or continuously in the body and stop and repair changes already in their beginnings. Possibly for the individual this would only mean to ingest a capsule containing the nanobots.

According to all we know, this is the only way a human being could stay young. Unfortunately, there would be a lot of technology involved, but so close to nature that we remain human beings.

One should also not conceal the fact that many dangerous possibilities of error and misuse are to be expected, as with any effective new technology. Nanotechnology would also have the possibility to destroy life and change the earth positively as well as negatively. Its development seems unstoppable. However, its use in cryonics would hardly contribute to misuse.

Literature

Best BP (2013b) Effects of temperature on preservation and restoration of cryonics patients. Cryonics Magazine (Institute Evidence-based Cryonics)

Carmichael ST (2006) Cellular and molecular mechanisms of neural repair after stroke: making waves. Ann Neurol 59:735-742

Chen J et al (1998) Induction of caspase-3-like protease may mediate delayed neuronal death in the hippocampus after transient cerebral ischemia. J Neurosci 18:4914-4928

Dancause N et al (2005) Extensive cortical rewiring after brain injury. J Neurosci 25:10167-10179

Drexler KE (1981) Molecular engineering. An approach in the development of general capabilities for molecular manipulation. Proc Natl Acad Sci USA 78:5275-5278

Drexler KE (1992) Nanosystems: molecular machinery, manufacturing, and computation. John Wiley & Sons Inc, New York

Drexler KE, Peterson C (1994) Experiment Zukunft. Addison-Wesley, Bonn

Fennimore AM et al (2003) Rotational actuators based on carbon nanotubes. Nature 424:408-410

Freitas RA Jr (2022) Cryostasis revival - the recovery of cryonics patients through nanomedicine. Alcor Life Extension Foundation, Scottsdale Arizona

Gertz HJ (1989) Neuronal plasticity in degenerative brain diseases. In Baltes M et al (eds) Successful aging conditions and variations. Huber, Bern, pp 250-253

Ideguchi M (2010) Murine embryonic stem cell-derived pyramidal neurons integrate into the cerebral cortex and appropriately project axons to subcortical targets. J Neuroscience 30:894-904

Kriegstein A, Alvarez-Buylla A (2009) The glial nature of embryonic and adult neural stem cells. Annu Rev Neurosci 32:149-184

Leunissen RL, Piatnek-Leunissen DA (1968) A device facilitating in situ freezing of rat heart with modified Wollenberger tongs. J Appl Physiol. 25:769-771

Mathwig K (2018) Molecular repair at physiological conditions? In Sames KH (ed) Applied Cryobiology Human Biostasis, vol. 2, ibidem, Stuttgart, pp 105-115

Mathwig K, Sames K (2013) Kryonik. In Sun MJ, Kabus A (eds) Reader zum Transhumanismus. Books on Demand, Norderstedt, Berlin, pp 113-129

McIntyre RL, Fahy GM (2018) Aldehyde stabilized cryopreservation (Reprint). In Sames KH (ed) Applied Human Cryobiology, vol. 2, ibidem, Stuttgart, pp 13-46

Merkle RC (1992) The technical feasibility of cryonics. Med Hypotheses 39:6-16

Merkle RC (1994) The Molecular Repair of the Brain. Cryonics (Alcor) 15:18-30

Merkle RC (1994a) The molecular repair of the brain. Cryonics (Alcor) 15:16-31

Merkle RC, Freitas RA (2008) A cryopreservation revival scenario using molecular nanotechnology. Cryonics 4th Quarter, Alcor

Merkle FT et al (2004) Radial glia give rise to adult neural stem cells in the subventricular zone. Proc Nat Acad Sci USA 10:17528-17532

Mezey E et al (2003) Transplanted bone marrow generates new neurons in human brains. Proc Natl Acad Sci 100:1364-1369

Mikula S (2016) Progress toward mammalian whole-brain cellular connectomics. Front Neuroanat 10:62. doi: 10.3389/fnana.2016.00062. eCollection 2016

Mikula S et al (2015) High-resolution whole-brain staining for electron microscopic circuit reconstruction. Nature Methods 12:541-546

Nudo RJ (2007) Postinfarct cortical plasticity and behavioral recovery. Stroke 38:840-845

Sames K (2000) Sterblich durch ein Gesetz der Natur? Frieling, Berlin

Sames K (ed) (2000a) Medizinische Regeneration and Tissue Engineering. Ecomed, Landsberg

Sames KH (2013) Organ differentiation and mortality. In Sames KH (ed) Applied Human Cryobiology, vol. 1, ibidem, Stuttgart, pp 125-144

Sames KH (2013b) General mechanisms of mortality and aging and their relation to cryonics. In Sames KH (ed) Applied Cryobiology, vol. 1, ibidem, Stuttgart, pp 145-169

Vita-More N et al (2015) Persistence of long-term memory in vitrified and revived Caenorhabditis elegans. Rejuvenation Res 18:458-463

Wu L et al (2008) Neural stem cells improve neuronal survival in cultured postmortem brain tissue from aged and Alzheimer patients. J Cell Mol Med 12:1611-1621

16 Outlook:
encouraging progress in cryonics research

Summary

Older and current findings and progress as well as new developments are summarized. They show that we are approaching the goal of preserving even large biological entities such as human organs in a viable state by cryopreservation. Research on organ transplantation is increasingly making cryopreservation of human organs a goal. This requires solving the problems that exist in cooling and heating large bodies, including human ones.

Despite all the problems, the overall cryonics project is making progress and finding new methods.
(in the following, references are omitted where it is a matter of mentioning facts discussed in detail above).

It is becoming increasingly apparent to researchers that animal life can survive at negative body temperatures down to cryogenic temperatures. In extreme cases, the structures of tiny creatures can be maintained in a viable state even without energy.

Interesting are the results according to which also small mammals can survive minus temperatures in the uppermost range both in hibernation and in cooling under laboratory conditions but in many cases without solidification of their entire body water. Warm-bloodedness is therefore not an absolute obstacle to cryopreservation.

Many cold-blooded animals routinely hibernate by partially freezing their body water.

Various small invertebrates even survive cryogenic temperatures, seemingly effortlessly. We find this ability in insects as well as the even smaller nematodes, rotifers and tardigrades. In the latter, the temperatures survived can be traced almost to absolute zero. This also shows that life can exist without energy in the form of structures. Nothing irreplaceable is lost during cooling.

The results signal that cryonics is not impossible in principle.

Previously, it was assumed that cryogenic temperatures were a prerequisite for safe storage of biological tissues in a viable state for thousands of years. Recently, however, it was shown that rotifers and nematodes apparently survive up to 20 000 years or longer at polar temperatures, i.e. above cryogenic temperatures

A number of findings in organ cryopreservation indicate that preservation of organs between 0° and -80°C is not as problematic as that at temperatures from -80°C down in the cryogenic temperature range. It remains to be tested whether large bodies can also be stably preserved at the higher temperatures and whether this is possible for similar time periods as for the polar tiny bodies. It is to be clarified whether the survival of the small animals is based on special properties.

Survival of oxygen-deprived cells is possible in the human body, and they can still be cultured and kept alive long-term after many hours of circulatory arrest when the human organs have failed. In particular, the finding that brain cells and their stem cells are still alive in large numbers hours after organ failure in humans, in slaughter animals, and in animal studies shows that cryonics can make sense on deceased individuals and creates starting points for delaying brain death. Cells, small tissue samples, and brain slices can be explanted and cryopreserved even after "death," i.e., after an animal has been killed. They can be revived after thawing. Connective tissue cells can be preserved for many hours after human "death." In a dying human body (after general organ failure), cooling and freeze protection are of course technically more difficult than in small, freshly removed objects.

Individual measures against changes in organ failure (i.e. after "death") are known. Such against the "no reflow" phenomenon, the complicated processes that clog the vessels when circulation stops, are now known to us, although still without comprehensive effect in resuscitating people. Obstruction of the cerebral vessels is a major cause of the inability to resuscitate people after a few minutes of circulatory arrest.

First success is also seen in preserving brain cells and protecting them against cell suicide when oxygen is depleted.

Advances in the organization of cryopreservation research have been increasing recently. The role of cryopreservation of human organs for transplantation medicine is increasingly realized. Therefore, organizationally, cryopreservation research in particular is currently developing in favor of organ transplantation.

An institute at the University of Minnesota is establishing an organ and tissue preservation center that will focus primarily on the warming of cryopreserved organs. There will then be two such institutes in the USA. In collaboration, the suspended animation will also be researched by the institutes. In 2017, a scientific article on organ cryopreservation was published by Giwa et al. which involved international collaborators from more than 40 renowned institutes, focusing on organ transplantation.

The first successes are becoming visible in the use of computer programs to cope with large amounts of data. The standardization of tests and computer programs in cryopreservation is making progress. They can be used, for example, to track the formation of ice crystals and the reduction or shrinkage of ice crystals by various substances during heating (Ampaw et al. 2020).

New antifreeze agents and new mixtures of antifreeze agents show progress in solving the most difficult problems. For example, the DP6 cryoprotective solution allows slower heating. Slow heating without the formation of ice crystals could bring cryonics one step closer to its realization. Increasing knowledge of the temperature-related properties of individual—as well as mixtures of different—cryoprotectants will enable a longer temperature equilibration at less damaging concentrations of each substance. Improvements in vitrifying solutions and new solutions have been tested (Fahy, Wowk 2015; -2021; Pollock et al. 2017; Wowk et al. 2018). Neutralizing the harmfulness of cryoprotectants by other cryoprotectants is increasingly known and exploited. One cryoprotectant can make others less harmful through general effects such as dilution (Fahy et al. 2004), reducing the harmfulness of each. But this can also occur through specific reactions between cryoprotective substances that depend on properties of the molecules. Further, one can choose cryoprotectants that react less with water so that the water layers around the biological molecules are not destroyed. It is possible to reduce the membrane-permeable cryoprotectants by using non-membrane-permeable and ice-blocking agents. Rapid cooling reduces the damaging effect and rapid heating hinders devitrification (Hawkins et al. 1985). A long-range goal is to explain the antifreeze effect comprehensively in terms of chemical properties of the individual cryoprotectants.

Initial successes have been achieved with directional freezing of organs, for example in the freezing of a whole rat leg. What is possible with this method will only become clear once directional vitrification has also been developed (Arav 2022).

For cryopreservation, often well-known facts play a decisive role. For example, perfusion of capillaries shortens the diffusion paths, because our most active cells are only nanometers away from the bloodstream. Therefore, diffusion of substances, including cryoprotectants, into the cells and temperature change are not as big a problem as one would expect from the size of the organ. Theoretically, the problems could start first with cracking or, if tissue is stored at a temperature near the glass transition (avoiding cracking temperatures), only with rewarming.

Cooling and cryoprotectants can, fluidity of solutions provided, therefore arrive very quickly in all cells in principle faster than a 1mm thick piece of tissue (without connection to a circuit) is passed through. Cooling and cryoprotectants may even reach the cells faster than in a cell culture that is not kept in a flow-through chamber. The difficulty lies more with temperatures below glass transition and, especially, heating. It remains to be determined how much the viscosity and cooling restrict flow in the blood vessels. Unfortunately, perfusion ends when a solid state is reached. The recent demonstration that a cryoprotectant solution penetrates membranes in seconds gives hope that antifreeze solutions can be provided in smaller concentrations and for shorter times than has been used in the past. It must then be cooled very quickly during a vitrification process. Heat exchange is similarly rapid during flow-through (perfusion).

If cryoprotectants could be kept in a liquid phase and circulated throughout the cooling process and during reheating, many problems should be solved.

Since the cooling and supply of cryoprotectant via the circulation can be applied below +10°C, where the harmful effects of the substances are already lower, organs of any size at the freezing point should already be saturated with antifreeze, so that there is protection against crystallization.

Harmful effects during cooling can be avoided relatively well up to the glass transition temperature. The impossibility of resuscitation emerges more in stress fractures and devitrification. Methods of rewarming could thus lead to a breakthrough. Electromagnetic heating via nanoparticles brought into capillaries is taking on more concrete forms. It makes heating of frozen tissues to the normal living state possible. Heating by focused ultrasound is also promising, having already been survived by nematode worms. Its application to larger objects appears feasible.

The good successes on organs and animals are — as mentioned — predominantly in the temperature range between 0- and -80°C. The difficulties below this temperature range could perhaps have to do with the formation of crystallization nuclei. It remains to be investigated where the critical temperature limit lies, which is responsible for the fact that only very small organs can be cooled to the temperature of liquid nitrogen. Although the survival of tiny animals at polar temperatures for millennia is not readily transferable to organs of more highly developed animals, it should be clarified once again how well the latter can survive at such temperatures.

Improvement of cell membrane permeability is possible and has been achieved (de Graaf et al. 2007). This helps to control the effects of rapid cooling and shortens perfusion times, which is necessary to avoid harmful effects of cryoprotectants.

The development of various innovative methods is foreseeable.

Isochoric cooling works with increased pressure. Temperature changes lead to changes in the volume of substances and liquids. This is particularly pronounced in aqueous solutions during evaporation and ice crystal formation. The latter lead to expansion, which can be reduced by increasing the pressure. In a pressure-resistant closed chamber, the volume cannot increase. In contrast, the pressure increases and prevents expansion and thus ice crystal formation. Earlier this method has been used in histochemistry. Thus, in pressure-resistant chambers, ice formation can be avoided with little or no use of antifreeze (isochoric cooling method). In this way, subcooling down to -20°C was achievable. With deeper cooling, the formation of ice crystals is possible, but is slowed down by the resulting pressure in a system that does not allow for expansion. In this case, biological samples contained in the solution remain ice-free. If one proceeds here following vitrification methods down to cryogenic temperatures, one can determine whether ice develops by reading the pressure. Vitrification seems to be possible, the effort is low, because no energy input and no complicated technique are needed to generate the pressure.

The great advantage, according to what has been said, would be the possibility to work without ice crystals and with reduced harmful effects.

How the pressure change affects living units is not yet fully decided, although nematodes have survived the procedure. Theoretically, large pressure-resistant containers can also be used. The damage-free duration of supercooled storage in such a system is not yet known (Miku s et al. 2016; Năstase et al. 2017; Rubinsky et al. -2005; -2015; 2015a; -2021; Taylor et al. 2019; Ukpai et al. 2017; Wan et al. 2018; Zhang et al. 2018).

According to Newton, the viscosity of solutions does not change when shear forces are applied. However, there are solutions such as blood (non-Newtonian fluids), which change their flow properties under shear stress (so blood can pass through the narrow hair vessels).

Taylor et al. (2019) report on still largely theoretical considerations of increasing during cooling the viscosity of cryoprotective solutions ice-free by shear forces until they resemble glass. This would avoid ice and allow cryoprotectant to be used at lower concentrations that are less harmful.

It would also be possible to use super magnetic particles in a liquid that can increase their viscosity in a magnetic field (rheomagnetic solutions).

The amazing results of suspended animation at body temperatures as low as 10°C suggest that one continues to look for ways to lower temperatures starting from suspended animation and test antifreeze agents for this task. An interesting question would have to be whether this would also be possible after organ failure. This could only be clarified in animal experiments, since resuscitation is the measure of this.

Equally interesting would be the question of whether methods can be developed that allow animals to be cooled more deeply during hibernation, which is not possible by simply lowering the temperature, perhaps because the cells themselves do not build up sufficient frost protection and only their environment cools down (see in Fuhr et al. 2013 chap. 2).

Overall, cooling to cryogenic temperatures and subsequent resuscitation in vertebrates or their organs is one of the next goals. There is encouraging methodological progress in this direction. Doubts about feasibility are fading as knowledge and capabilities increase. Changes in organs due to aging and disease remain, for now, a long-term problem on the way to extending human lifespan. In younger people, however, this problem would initially be limited.

The most promising method actually in discussion is organ cryopreservation by vitrification and nanowarming using organ perfusion with solutions containing nanoparticles. The particles remain in the blood vessels during vitrification and when they are warmed up, the vitrified solution in the vessels is liquified. The further steps of warming and washout can then be performed by perfusion. In principle the method — successful in rat kidneys — has the potential to be adapted to the much larger human organs.

Literature

Ampaw AA et al (2021) Use of ice recrystallization inhibition assays to screen for compounds that inhibit ice recrystallization. Methods Mol Biol. 2021;2180:271-283. doi: 10.1007/978-1-0716-0783-1_9. PMID: 32797415.

Arav A (2022) Cryopreservation by directional freezing and vitrification focusing on large tissues and organs Cells 11:1072; https://doi.org/10.3390/cells11071072

De Graaf IA et al (2007) Cryopreservation of rat precision-cut liver and kidney slices by rapid freezing and vitrification. Cryobiology 54:1-12

Fahy GM, Wowk B (2015) Principles of cryopreservation by vitrification. In Wolkers WF, Oldenhof H (eds) Cryopreservation and freeze-drying protocols. Methods Mol Biol 1257, Springer Protocols Humana Press, Totowa, pp 21-82

Fahy GM, Wowk B (2021) Principles of ice-free cryopreservation by vitrification. In Wolkers WF, Oldenhof H (eds) Cryopreservation and freeze-drying protocols. 4th ed. Methods Mol. Biol 2180, Springer Protocols Humana Press, Totowa, pp 27-97

Fuhr G et al (2013) Unterbrochenes Leben? Naturwissenschaftliche und rechtliche Betrachtung der Kryokonservierung von Menschen. Fraunhofer, Stuttgart

Giwa S et al (2017) The promise of organ and tissue preservation to transform medicine. Nat Biotechnol 3:530-542

Hawkins HE et al (1985) The influence of cooling rate and warming rate on the response of renal cortical slices frozen to -40 degrees C in the presence of 2.1 M cryoprotectant (ethylene glycol, glycerol, or dimethyl sulfoxide). Cryobiology 22:378-384

Mikus H et al (2016) The nematode Caenorhabditis elegans survives subfreezing temperatures in an isochoric system. Biochem Biophys Res Commun 477:401-440

Năstase G et al (2017) Isochoric and isobaric freezing of fish muscle. Biochem Biophys Res Commun 485:279-283

Pollock K et al (2017) Improved post-thaw function and epigenetic changes in mesenchyme stromal cells cryopreserved using multicomponent osmolyte solutions. Stem Cells Dev 26:828-842

Rubinsky B (2005) The thermodynamic principles of isochoric cryopreservation. Cryobiology 50 121-138

Rubinsky B (2015) Biological matter in isochoric systems. Cryobiology 71:172-178

Rubinsky B et al (2015a) From ice in the veins, through unfrozen fish and frozen frogs to isochoric preservation, ad astra. Cryobiology 71:167-168

Rubinsky B (2021) Mass transfer into biological matter using isochoric freezing. Cryobiology 100:212-215

Taylor MJ et al (2019) New approaches to cryopreservation of cells, tissues, and organs. Transfus Med Hemother 46:197-215

Ukpai G et al (2017) Pressure in isochoric systems containing aqueous solutions at subzero centigrade temperatures. PLoS One 12 Aug(8):e0183353

Wan L et al (2018) Preservation of rat hearts in subfreezing temperature isochoric conditions to -8 °C and 78 MPa. Biochem Biophys Res Commun 49:852-857

Wowk B et al (2018) Vitrification tendency and stability of DP6-based vitrification solutions for complex tissue cryopreservation. Cryobiology 82:70-77

Zhang Y et al (2018) Isochoric vitrification: an experimental study to establish proof of concept. Cryobiology 83:48-55

Index

Aging 124
Air bubbles 216
Amides 73
Alcoholic groups 69
Animaliculi "Diertgens" 180
Anticoagulants 221
Anticoagulation 256
Antifreeze agents 68, 105
Antifreeze heterothermic vertebrates 177
Antifreeze in our bodies 78, 105
Antifreeze solutions harmful concentrations 122
Antifreeze solutions, measure of damaging effect: qv* 84
Antifreeze substances 69
Antifreeze substances in animals 177, 178
Antifreeze proteins, water molecules 180
Antioxidants 216
Antioxidants 259
Aorta 214
Arachidonic acid 197
Arterial ring, brain (circulus Willisii) 214
Artificial bone 173
Artificial cartilage 174
Artificial esophagi 173
Artificial skeletal muscle 173
Artificial skin 173
Artificial tissue, cryopreservation 172-174
ATP, xanthine 202

Beetle Anatolica polita 179
Beetle Cucujus clavipes puniceus 179
Beetle Dendroides Canadensis 179
Beetle Perostichus brevicornis 179
Beetle Upis ceramboides 179
Biostasis 21
Blood-brain barrier 83, 120
Blood-brain barrier, damage 202
Blood circuit, artificial 213
Blood, closed artificial circuit 213
Blood exchange 212-213
Blood open artificial circuit 213
Blood vessel occlusion ("no reflow") 203-204, 231
Blood vessels, cryopreserved 152
Body kept youthful 273
Bone, cryopreservation 156
Bound water 103
Brain 169-172
Brains, cat -20°C 171
Brain cells, plasticity 268
Brain cells recovering after death 240
Brain cells survival 235, 240
Brain death, loss of functions or loss of cells 231
Brain neurons replacement by astrocytes 267

Brain neurons, replacement by stem cells 257
Brains of pigs, slaughtered, kept alive 4 hours 237
Brain repaired in its personal form 272
Brain survival, state of science 244
Brain swelling 120
Brimstone butterfly 179
Bypass vessels 214

C. elegans, cryogenic temperatures, memory 268
Capillaries 123
Capillaries, from to cells 127
Cardiac arrest 36
Cardiac surgery 43
Cardiac tissue samples -196°C 151
Cardiopulmonary resuscitation device 212
Carrier solutions 218
Carotid arteries, perfusion 214
Carrier solutions improve perfusion 219
Carrier solution M-RPS-2 219
Cartilage 77, 78, 156
Caspase-3 inhiitor ZDEVD-FMK 259
Cell death, rapid, necrosis 202, 232
Cell death, stop of circulation 233
Cell membrane K/Na ratio 130
Cell membrane permeability improvement 281

Cell membrane protection 95, 96
Cell membranes 56
Cell shrinkage 70
Cell suicide apoptosis 202, 232
Cell suicide intervention 259
Cell survival 68
Cell types 57
Cell volume, critical minimum 54, 55
Cells arterial inner wall, shrinkage 222
Cells, lethal effects on 82
Cells outlive oxygen deprivation 243, 269
Chemical fixation 96
Chemical fixation, collapse artifacts 97
Chilling injury 38, 128-131
Chondroitin sulfate 77
Clathrates 110
Cold shock 38, 128, 129
Colligative interference 68
Colligative qualities 68, 70
Computer programs (and AI) 98
Cooling as soon as possible 211, 253-255
Cooling, isochoric 281
Cooling, large-body 128
Cooling rates 54-56
Cooling rate, optimal 56
Cooling rates critical 73
Cooling slow 54, 71
Cooling without antifreeze (freezing) 213

Coordinates of all body molecules 273
Cornea, vitrification 149-151
Cortisone, related drugs 216
Cracks (freeze fractures) 123, 124
Cracks during heating 132
Cracking, reduction 218
Crackphone 125
Cryonics 21
Cryonics, functioning 22
Cryonics, premature 136
Cryonics scenario 25
Cryopreservation in nature 176-181
Cryopreservation in organ transplantation 279
Cryoprotectant, addition and dilution of 81, 82
Cryoprotectant, concentration of 54, 68
Cryoprotectant DP6 solution 86
Cryoprotectant flow to the brain 214
Cryoprotectants, cells, take up 222
Cryoprotectants, glass transition temperature 99
Cryoprotectants, harmful effect of 79, 84, 100
Cryoprotectants, non-specific toxicity/specific toxicity 80
Cryoprotectants readiness to form glass 100
Cryoprotectants reducing harmfulness of others 85
Cryoprotectants which protect cell membranes 95, 96

Cryoprotectants with large molecules 72
Cryoprotectant uptake, perfusion 127
Cryoprotective agents, penetrating (CPAs) 69
Cryostasis 21
Crystal growth, speed of cooling 102
Crystal growth, high viscosity 102
Crystallization of water 42
Crystals, large 134
Crystallization nuclei 50
Cytoskeleton 84

Death 227-245
Death of brain 229
Deep-cooling influence (of) numerous factors 97
Depolarization 194, 195
Deprenyl 260
Devitrification 131, 133
Dextran 224
Diffusion, Fick's law of 53
Dilution 67
Dilution, blood (hemodilution) 255
DMSO (Dimethyl sulfoxide) 68, 70
Donor organs 27
DP6 (cryoprotective) solution 127
Drowning in icy water 230
Drugs during cooling 215
Drugs given before organ failure 211

Dry ice 217
Dying/decay, gradual (process) 230

ECMO (extracorporeal membrane oxygenation) 217
Electroencephalogram (EEG) 228
Electromagnetic heating 135
Electromagnetic heating nanoparticles 134
Embryo cryopreservation 146, 147
Embryo, human cryopreservation 147
Energy for distribution of ions 195
Energy reserves 196
Energy store ATP, loss of 198
Equilibrium temperature 69
ESA 30
Ethylene glycol 69, 95, 222
Excitotoxicity 198, 232

Femoral artery 213
Fertility preservation 147
Fertilization in vitro 147
Fingers, human cryopreserved, transplanted 154
Flow through capillaries 215
Free radicals 197, 201-203
Free radical scavenger NXY-059 259
Freeze fractures of smaller samples 125
Freezer computer-controlled 217
Freezing, directional 60

Freezing point 68
Freezing point, lowering 70
Frogs 177-178
Frost-resistant 52

Gene expression analyses 80
Glass 58
Glass formation 59
Glass transition temperature 59, 72, 99, 102, 103
Glass transition temperature, additives 103
Glycerol 74, 79, 102, 103
Ground substance (matrix) 81

Hansters below 0°C 174-175
Hearts, 27-36 kg pigs -3°C 152
Heart-lung machine, suitcase 217
Heart muscle cells, cryopreserved 144
Hearts, small -45°C 151
Heart valves, cryopreserved 151
Heat exchange, capillaries 127
Heat capacity of water 131
Heating, critical rates 73, 132
Heating rate 60
Heparin (Hep) 76, 215, 256-257
HES (hydroxyethyl starch) 72, 75, 220
Heteronuclei 51
Hibernation 181-182
High processing capacity, methods with 80
Homonuclei 51
Hyaluronic acid 77

Hydrogen bonds 68
Hydrogen sulfide (H2S) 38
Hydrophobic properties 68
Hydroxyl radicals 201
Hypothermia 51

Ice bath 212
Ice blockers 105-108, 216
Ice cream 104
Ice crystal axes 107
Ice crystallization during re-warming 106
Ice crystals 54
Ice crystals, damaging 57
Imaging techniques 82
Inflammatory vascular responses 202
Intestine, dog -196°C 154
Ischemia, cold 37
Ischemia, warm 37
Isotonic 53
Ice crystallization during re-warming 106

Kidney 153-154
Kidney, rabbit -130°C transplanted 153-154
K/Na ratio 130

Leech Ozobranchus jantseanus 180
Legs rats -140°C 154
Liver 155-156
Lividity 229
Lung, dog below 0° 155

M22 (cryoprotectant) 86, 101
Mammalian body temperatures below 0°C 177
Mannitol 224
Mealworm Tenebrio Molitor 179
Membrane potential 194
Membrane stability, dexamethasone 220
Melatomin, Hydroxyl radicals 260
Melting 58
Melting point 68
Melting temperature 51
Memory, consciousness, individuality 241
Methoxyl groups 70
Methylene blue, nitric oxide 202
Molecular movements 42
Moss lawn 180
Müller cells retina, cryopreserved 144

Nanobatteries 272
Nano jump 273
Nano particles, electromagnetically warmed 271
Nanorobots 271
Nanotube cables 272
NASA 30
Nematode Caenorhabditis elegans 180
Nematode turbatrix aceti 180
Nematodes in 30,000-year-old strata 176
Nervous tissue, cryopreserved 169

Neuron survival cocultivation 258
Neuron survival, Insulin 260
Neurosurgery 43
Nimodipine 261
Nitric oxide 201
Nitrogen vapor 217
NMDA channel 196
No-reflow 200
Nucleating proteins 52

Obstacles for cryonics 122
Oocytes vitrified 144
Organ, large 126
Organoids 173
Organ protection solutions 218
Organs 80
Osmotic pressure 54, 68
Osmotic pressure, changes in 82
Osmotic pressure within the capillaries 83
Ovarian tissue, Cryopreservation 148
Ovarian tissue reimplantation 148
Ovaries cryopreserved, sheep 149
Oxygen administration 212
Oxygen deprivation 39, 196
Oxygen, harmful reoxygenation 199
Oxygen, reactive products 201
Oxygen reperfusion 215
Oxygen, residual 39

Pancreas, cryopreserved 156
Pancreatic islets, cryopreserved 153
Patient cooled and stabilized 213, 217
Patient stored 218
Penetrating cryoprotective agents (CPAs) 69
Perfluorocarbon 40
Permeability of membranes 53, 56, 69, 70, 83
Peroxinitrite 201
Plasminogen activator 256-257
Platelets, chilling injury 129
Polyethylene glycol (PEG) 72
Polymers (cryoprotective) 72
Polyols or diols 69
Polysaccharides 75
Polyvinylpyrrolidone (PVP) 72
Pronouncement of death 227
Proteins, freezing point 178
Proteoglycans 72, 76-78, 101

Q10 Coenzyme 259

Radiation 41
Recrystallization on rewarming 131, 134
Red blood cells, cryopreservation 145
Reoxygenation 201-203
Reoxygenation in old bodies 202
Repair using nanotechnology 270
Reperfusion syndrome 199, 203
Reptiles supercooling 177

Resuscitation 22, 218, 228
Rewarming 269
Rewarming, institutes in USA 279
Rigor mortis 229
Rotifers 180
Rotifers, permafrost 24 000 years 176

Shear forces 282
Shipping in dry ice 132
Silicone bath 217
Smooth muscle -21°C 149
sodium/potassium pump 194
Space travel 29
Specimens, large 52
Sperm, cryopreservation 144
Spleen, canine, freezing transplantation 155
Standby team 211
Stem cells 243
Stem cells, cryopreserved 143
Stem cells, dormant 234
Stem cells, forebrain 234
Stem cells, repair 269
Storage just below -138- 140°C 124
Streptokinase 256-257
Sucrose 74, 105, 108
Sugars 74
Supercooled 39, 51, 67
Superoxide 201
Superoxide dismutase (SOD), chilling injury 129
Suspended animation, EPR 38, 199

Swelling, aqueous, (of) tissue 83, 223
Swelling/shrinking (of cells) 53, 81

Temperature, crystallization nuclei 132
Temperature, growth of crystals 132
Testicular tissue, immature, cryopreservation 148-149
Thawing 68
Thawing by microwaves 135
Thermal conductivity 125
Thermal hysteresis 179
Thermodynamic equilibrium 55
Thermodynamic stress 123
Tissue components completely replaced 272
Tissue samples 28
Tissue structure preserved, after death 269
Tooth -150°C, transplanted 156
Transmitter 195
Transport time 217
Trehalose 72, 74, 75, 105, 108
Trehalose, chilling injury 130

Ultrasound focused heating 135
Ureter, canine cryopreservation, transplantation 155
Uteruses, pigs -130°C 154

VS55 (cryoprotectant) 86
Vena cava, superior 214
Viaspan 219

Viscosity, increase by pressure 101

Viscosity of membranes 56

Vitamin E 257

Glass formation (Vitrification) 59

Vitrification, directional 61

Vitrification, disadvantages 123

Vitrification if possible 216

Vitrification, nano-warming 283

Vitrification solution 103

Vitrification, strong concentration of cryoprotectants 99

Vtrificationi, viscosity 99

Vitrification within the cell 72

Vitrify 2 L 102

VM1 (cryoprotectant) 86

VM1 damaging 108

VM1 vitrifies stably 108

Washing out of the blood 212, 215, 219, 255

Water bear (Tardigrada) 180-181

Water-binding capacity 68, 70

White blood cell adhesion 202